SpringerBriefs in Meteorology

SpringerBriefs in Meteorology present concise summaries of cutting-edge research and practical applications. The series focuses on all aspects of meteorology including, but not exclusive to; tornadoes, thunderstorms, lightning, hail, rainfall, fog, extratropical and tropical cyclones, forecasting, snowfalls and blizzards, dust storms, clouds. The series also presents research and information on meteorological technologies, meteorological applications, meteorological forecasting and meteorological impacts (reports of notable worldwide weather events). Featuring compact volumes of 50–125 pages (approx. 20,000–70,000 words), the series covers a range of content from professional to academic such as: a timely reports of state-of-the art analytical techniques, literature reviews, in-depth case studies, bridges between new research results, snapshots of hot and/or emerging topics. Author Benefits: Books in this series will be published as part of Springer's eBook collection, with millions of users worldwide. In addition, Briefs will be available for individual print and electronic purchase. SpringerBriefs books are characterized by fast, global electronic dissemination and standard publishing contracts. Books in the program will benefit from easy-to-use manuscript preparation and formatting guidelines, and expedited production schedules. Both solicited and unsolicited manuscripts are considered for publication in this series. Projects will be submitted for editorial review by editorial advisory boards and/or publishing editors. For a proposal document please contact your Publisher, Dr. Robert K. Doe (robert.doe@springer.com).

More information about this series at http://www.springer.com/series/13553

Wayan Suparta · Kemal Maulana Alhasa

Modeling of Tropospheric Delays Using ANFIS

 Springer

Wayan Suparta
Space Science Centre (ANGKASA)
Universiti Kebangsaan Malaysia
Bangi
Malaysia

Kemal Maulana Alhasa
Space Science Centre (ANGKASA)
Universiti Kebangsaan Malaysia
Bangi
Malaysia

ISSN 2199-9112   ISSN 2199-9120   (electronic)
SpringerBriefs in Meteorology
ISBN 978-3-319-28435-4   ISBN 978-3-319-28437-8   (eBook)
DOI 10.1007/978-3-319-28437-8

Library of Congress Control Number: 2015958860

Printed on acid-free paper

This Springer imprint is published by SpringerNature
The registered company is Springer International Publishing AG Switzerland

Preface

The Global Navigation Satellite Systems (GNSS) technology was implemented in the field of meteorology in the early 1990s, where scientists successfully developed a technique for determining water vapor in the troposphere by exploiting the signal errors during propagation from satellite to a receiver. One of the main sources of error in the GNSS and its impact, which are crucial in efforts to improve the accuracy of positioning, is the zenith tropospheric delay (ZTD). If ZTD is produced with fine temporal and spatial resolution, the value and variability can be applied to meteorological studies. In other words, ZTD in this case is the main parameter that plays an important role in determining the parameters of water vapor in the atmosphere.

In the advantages of GNSS such as GPS technology for atmospheric research applications, it was found that this method or other methods do not always excel in all circumstances. For example, GPS data is not always available for a full 24-h period, especially for a remote location or a strategic area where the GPS receiver is not installed, while the accessibility and accurate estimation of this parameter is necessary. Hence, we look at a different approach that is cost-effective and robust in retrieving the value of ZTD by applying a soft computing technique such as adaptive neuro-fuzzy inference system (ANFIS) as a new alternative. There are various approaches by other soft computing techniques, such as genetic algorithms (GA), artificial neural network (ANN), fuzzy logic model (FLM), and particle swarm techniques. However, ANFIS was selected as it is emerging as a potential and robust optimization tool in recent years. ANFIS is a method that combines ANNs and a fuzzy inference system. In this technique, a fuzzy clustering algorithm is adopted to enhance the performance of the models, which is able to minimize the number of membership functions and rules for better efficiency of the models.

To investigate the accuracy of the ZTD models developed, a combination of the surface pressure (P), temperature (T), and relative humidity (H) is analyzed to obtain the best estimation of ZTD. The results demonstrate that ANFIS models with three inputs network (P, T, and H) agree very well with the ZTD obtained from GPS. Finally, the three-input network is selected for developing the ZTD predictive models. To perform the ZTD model, five selected stations over Antarctica and three selected stations in regions of Malaysia and Singapore were used to examine the applicability of ANFIS. The ZTD prediction is performed from one to eight-step ahead for Antarctica region and from one to fifteen-step ahead for the equatorial region. The results demonstrate that ANFIS is capable of predicting ZTD with high accuracy.

This book is prepared to help students, lecturers, engineers, geodesists, meteorologists and climatologists, or practitioners to develop new knowledge in applications of the soft computing technique. The example application of soft computing presented here is for estimation and prediction of tropospheric delays. On the other hand, the ZTD parameter obtained from the models or measurements needs to be converted into precipitable water vapor (PWV) to make it more useful for weather forecasting, analysis of atmospheric hazards such as tropical storms, flash floods, landslides, and earthquakes, as well as for climate change studies.

In this book, readers are presented with a detailed theoretical background of ANN and ANFIS in Chap. 2. Chapter 3 describes the modeling of tropospheric delay and mapping functions from GPS observations. Chapter 4 presents the implementation of ANFIS model for estimation of ZTD. In this chapter, the ZTD value from both ANFIS models and GPS measurements for Antarctica and the equatorial regions is compared. The accuracy of each ZTD model for each input used in the training, testing, and validation is comprehensively elucidated, and finally in Chap. 5 the reader is introduced to the prospect of ZTD estimation using ANFIS, which focuses on how to predict the ZTD value using the surface meteorological data as input. With its simple writing style, it is hoped that this book will provide complete knowledge to readers on application of soft computing, in particular, for meteorological applications and processes involved in the method of observation, data processing, analysis, and methods of data interpretation.

In addition to the content of the book, the authors are particularly grateful to agencies such as Scripps Orbit and Permanent Array Center (SOPAC) for archiving GPS data, Crustal Dynamics Data Information System (CDDIS) NASA for archiving the ZPD data, the Australian Antarctic Division (AAD) and the British Antarctic Survey (BAS) for the surface meteorological data, and ANZ and NIWA for the Scott Base data where some data used in the analysis is their contribution. Finally, the author also expresses his gratitude to the publishers for agreeing to publish this book.

Wayan Suparta

Contents

List of Figures

List of Tables

Chapter 1
Introduction

Abstract Tropospheric delay is one of the atmospheric quantities, which today plays a crucial role in meteorological studies and weather forecasts as well as the positioning accuracy in altitude determination. The current state analysis revealed that the tropospheric delay is retrieved from the *Global Navigation Satellite System* (*GNSS*) receivers, which is known as the total troposphere zenith path delay (ZPD), or often referred to as zenith tropospheric delay (ZTD). In this method, GNSS (e.g., Global Positioning System (GPS) for simplicity) satellite sends electromagnetic signals through the atmosphere to a receiver on the ground at a fixed location. The electromagnetic signals are delayed due to the high amount of dry gas and water vapor in the troposphere layer. Total delay in the GPS signals is measured, and ZPD is obtained from a summation of dry and wet components (Hofmann-Wellenhof in Atmospheric effect on the global positioning system, theory and practice. Springer, Berlin, 2001). Because of the importance of ZPD data, Byun and Bar-Sever (Adv Eng Soft 31:312–321, 2009) have updated the legacy of new ZPD products for all available IGS stations with improving consistency of time series, which enhance climate studies.

Keywords Tropospheric delays · GPS · Artificial intelligence · Numerical modeling · Meteorological applications

The significance of zenith path delay (ZPD) data, such as is to improve numerical weather prediction (NWP) models and weather forecasting, where the accuracy of its value is converted into the amount of water vapor in the atmosphere, which is the so-called precipitable water vapor (PWV). PWV is one indicator of the critical components of our atmosphere, where their distribution and content are critical variables that can explain the evolution of various physical processes in the atmosphere. The role of water vapor as noted by a variety of authors (Dai et al. 2009; Ware et al. 2001; Valeo et al. 2005; Choy et al. 2013) is not only important in the formation process of clouds and aerosols and the chemistry of the lower atmosphere, but also closely related to the formation of rain, snow, and thunderstorms. Thus, the information about the quantity of water vapor is very important because it is useful for research, mainly in the field of hydrology, climate, and

© The Author(s) 2016

W. Suparta and K.M. Alhasa, *Modeling of Tropospheric Delays Using ANFIS*, SpringerBriefs in Meteorology, DOI 10.1007/978-3-319-28437-8_1

meteorology. These two parameters that are derived from Global Positioning System (GPS) signals should be available with high temporal and spatial resolutions. The availability of this data with high accuracy and accessible is necessary, especially lately there has been a striking increase in climate change which must be mitigated and adapted.

Since 1990, GPS has become a versatile tool with a low cost for remote sensing of ZPD (e.g., Bevis et al. 1992; Bar-Sever et al. 1998; Gendt 1998; Pottiaux and Warnant 2002; Suparta et al. 2008; Vázquez and Grejner-Brzezinska 2013). The data obtained has an advantage with high spatial and temporal resolutions and the cost of receiver maintenance is low. However, the performance of the most ground-based GPS systems is still lacking in providing continuous data for a full 24-h period, particularly in remote areas, and in some cases there is no GPS receiver installed. This condition may affect the achievement of ZPD data for retrieving the water vapor data with high spatial and temporal resolutions and thus it can affect the accuracy of weather forecasting system. Here, we propose artificial intelligence (AI) method, which is progressively and successfully applied to model the non-linear systems in engineering and scientific applications. One part of a popular AI is the artificial neural networks (ANN). ANN has been adopted by scientists to model complex nonlinear systems and has also been widely applied to problems in the field of climate forecasting. In the field of hydrology, Luk et al. (2001) employed the three methods of ANN, i.e., Multi-Layer Feed-Forward Network (MLFN), Elman neural networks and Time Delay Neural Network (DNN) to develop a forecasting model of rain. Their results for the entire neural network can make a reasonable prediction of rainfall for one-step forward (15-min). Bodri and Cermak (2000) have also successfully modeled the rainfall and rainfall forecasting. Furthermore, Kisi (2007) has used ANN to predict the flow of a river in the United States. He adopted three learning algorithms: back propagation, conjugate gradient, and Levenberg–Marquardt (LM) to construct the river flow forecasting model. The results showed that ANN model with LM learning algorithm is the best to forecast short-term river flow compared to the two other models.

The advantage of ANN over the traditional method is that it does not need to know about the physical relationship for systematically converting an input to output. The ANN can adapt itself to self-organize its structure, when the sample input training is presented. Although ANN offers several advantages, they still have a number of limitations such as to reach convergence rate is slow, including the potential trapped in a local minima and difficulty in selecting the appropriate architecture (Suykens 2001; Dogan et al. 2010).

To overcome this limitation and enhance the capability and accuracy, the integration of neural networks and fuzzy logic should be conducted to produce a new method, which is neural-fuzzy. This integration has the ability to transform the qualitative aspects of human knowledge and reasoning into the exact quantitative analysis. Here, the new technique namely neuro-fuzzy system (Rajasekaran and Pai 2003) can take the potential benefits of both models into a single framework. Neuro-fuzzy system can eliminate the basic problem of fuzzy modeling, where it using the learning capability of ANN for the extraction automatic fuzzy If-Then rule and parameter optimization. In addition, this method can utilize the linguistic information from a human expert knowledge as well as data measured during the development of the model (Nayak et al. 2004). This method can learn independently and adapt itself to its environment. In this paradigm, one of the significant developments from the integration of neural networks and fuzzy logic is adaptive neuro-fuzzy inference system (ANFIS). ANFIS model has been shown to be powerful in various applications such as a power system dynamic load (Altug et al. 1999; Djukanovic et al. 1997), short wind prediction (Negnevitsky et al. 2007; Xia et al. 2010) and time series simulation (Soto et al. 2013).

This study attempted to apply ANFIS modification network for development of tropospheric delays to estimate the ZPD value over the Antarctica and Malaysia regions. Tropospheric delay is a function of the satellite elevation angle and the geographic position of the receiver, and is dependent on the atmospheric pressure, temperature, and water vapor characteristics. Therefore, to build the model, we only considered the surface meteorological data (Pressure (P), Temperature (T), and Relative humidity (H)) to be used as input data network and ZPD retrieved from ground-based GPS is used as the target outputs and model validation. In addition, the multi-layer perceptron (MLP) and multiple linear regressions (MLR) methods were selected as a comparison model to test the ability of ANFIS model in estimating the value of ZPD. In other words, ANFIS model is developed to estimate ZPD value, and later without the need to use a GPS receiver to obtain ZPD data. Indeed, the selection of input variables is the most important step because it has an impact on the determination of the network architecture of ANFIS models (Nourani and Komasi 2013). In addition, the amount of data used for training will affect the ability of the model to capture various characteristics of ZPD in the concerned area. Finally, the ZPD predictive model from one-step to eight-step ahead is developed from three input networks (P, T, and H). The accuracy of the predictive model will be presented in the subsequent chapters.

References

Altug S, Chow MY, Trussell HJ (1999) Fuzzy inference systems implemented on neural architectures for motor fault detection and diagnosis. IEEE T Ind Electron 46(6):1069–1079

Bar-Sever YE, Kroger PM, Borjesson JA (1998) Estimating horizontal gradients of tropospheric path delay with a single GPS receiver. J Geophys Res-Sol Earth 103(B3):5019–5036

Bevis M, Businger S, Herring TA, Rocken C, Anthes RA, Ware RH (1992) GPS-meteorological: remote sensing of atmospheric water vapor using the global positioning system. J Geophys Res-Atmos 97:15787–15801

Bodri L, Cermak V (2000) Prediction of extreme precipitation using a neural network: application to summer flood occurrence in Moravia. Adv Eng Soft 31:312–321

Byun SH, Bar-Sever YE (2009) A new type of troposphere zenith path delay product of the international GNSS service. J Geod 83(3–4):1–7

Choy S, Wang C, Zhang K, Kuleshov Y (2013) GPS sensing of precipitable water vapour during the March 2010 Melbourne storm. Adv Space Res 52(9):1688–1699

Dai J, Wang Y, Chen L, Tao L, Gu J, Wang J, Xu X, Lin H, Gu Y (2009) A comparison of
    lightning activity and convective indices over some monsoon-prone areas of China. Atmos Res
    91(2–4):438–452
Djukanovic MB, Calovic MS, Vesovic BV, Sobajic DJ (1997) Neuro-fuzzy controller of low head
    hydropower plants using adaptive network based fuzzy inference. IEEE T Energy Conver
    12(4):375–381
Dogan E, Gumrukcuoglu M, Sandalci M, Opan M (2010) Modelling of evaporation from the
    reservoir of Yuvacik dam using adaptive neuro-fuzzy inference systems. Eng Appl Artif Intel
    23(6):961–967
Gendt G (1998) IGS combination of tropospheric estimates—experience from pilot experiment. In:
    Dow JM, Kouba J, Springer T (eds) Proceedings of 1998 IGS analysis center workshop. IGS
    Central Bureau, Jet Propulsion Laboratory, Pasadena, pp 205–216
Hofmann-Wellenhof B, Lichtenegger H, Collins J (eds) (2001) Atmospheric effect on the global
    positioning system, theory and practice. Springer, Berlin
Kisi O (2007) Streamflow forecasting using different artificial neural network algorithms. J Hydrol
    Eng 12(5):532–539
Luk KC, Ball JE, Sharma A (2001) An application of artificial neural networks for rainfall
    forecasting. Math Compute Model 33:683–693
Nayak PC, Sudheer KP, Rangan DM, Ramasastri KS (2004) A neuro-fuzzy computing technique
    for modeling hydrological time series. J Hydrol 291(1):52–66
Negnevitsky M, Johnson, PL, Santoso S (2007) Short term wind power forecasting using hybrid
    intelligent systems. In Power engineering society general meeting 2007 IEEE. doi:10.1109/
    PES.2007.385453
Nourani V, Komasi M (2013) A geomorphology-based ANFIS model for multi-station modeling
    of rainfall–runoff process. J Hydrol 490:41–55
Pottiaux E, Warnant R (2002) First comparisons of precipitable water vapor estimation using GPS
    and water vapor radiometers at the Royal Observatory of Belgium. GPS Solut 6(1–2):11–17
Rajasekaran S, Pai GV (2003) Neural networks, fuzzy logic and genetic algorithm: synthesis and
    applications (with cd). PHI Learning Pvt Ltd
Soto J, Melin P, Castillo O (2013) A new approach for time series prediction using ensembles of
    ANFIS models with interval type-2 and type-1 fuzzy integrators. In Computational intelligence
    for financial engineering & economics (CIFEr), pp 68–73
Suparta W, Abdul Rashid ZA, Mohd Ali MA, Yatim B, Fraser GJ (2008) Observations of antarctic
    precipitable water vapor and its response to the solar activity based on GPS sensing. J Atmos
    Sol-Terr Phys 70:1419–1447
Suykens JAK (2001) Nonlinear modeling and support vector machine. In Instrumentation and
    measurement technology conference 2001 (IMTC 2001), pp 287–294
Valeo C, Skone SH, Ho CLI, Poon SKM, Shrestha SM (2005) Estimating snow evaporation with
    GPS derived precipitable water vapour. J Hydrol 307(1–4):196–203
Vázquez BGE, Grejner-Brzezinska DA (2013) GPS-PWV estimation and validation with
    radiosonde data and numerical weather prediction model in Antarctica. GPS Solut 17(1):22–39
Ware RH, Fulker DW, Stein SA, Anderson DN, Avery SK, Clark RD, Droegemeir KK,
    Kuettner JP, Minster JB, Sorooshian S (2001) Real-time national GPS network for atmospheric
    research and education. J Atmos Sol-Terr Phys 63(12):1315–1330
Xia J, Zhao P, Dai Y (2010) Neuro-fuzzy networks for short-term wind power forecasting. In 2010
    International conference on power system technology (POWERCON). doi:10.1109/
    POWERCON.2010.5666697

Chapter 2
Adaptive Neuro-Fuzzy Interference System

Abstract This chapter explains in detail the theoretical background of Artificial Neural Network (ANN) and Adaptive Neuro-Fuzzy Inference System (ANFIS). The detailed explanation of this method will highlight its importance in the estimation of ZTD model.

Keywords Artificial neural network · ANFIS · Fuzzy inference system · Hybrid learning algorithm · Backpropagation

2.1 Artificial Neural Networks

Generally, an artificial neural network (ANN) is a system developed for information processing, where it has a similar way with the characteristics of biological neural systems. It was developed based on the human brain, which is capable of processing information, which are complex, nonlinear, and being able to work in parallel, distributed, and local processing and adaptation. ANN is designed to resemble the brain systems such as the construction of architectural structures, learning techniques, and operating techniques. This is the reason that ANN has been widely adopted by scientists because of its accuracy and its ability to develop complex nonlinear models and is used to solve a wide variety of tasks, especially in the field of climate and weather. This section will discuss the capabilities of ANN such as neurons modeling, architecture, and its learning process.

2.1.1 Neuron Modeling

In the human brain, there are neurons that are interconnected to one another. These neurons act as a tool that can perform processing of information of human senses. Haykin (2009) described that a biological neuron consists of a cell body, where

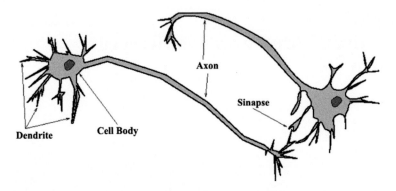

Fig. 2.1 Schematic diagrams of biological neurons (Haykin 2009)

conditions are covered by the cell membrane (Fig. 2.1). Each cell has branches called dendrites. Dendritic play a role in receiving the information into the cells of the body through the axon.

The axon is a long single fiber that can carry the signal from the cell body toward the neuron—the next neuron. The meeting point between neurons with the next neuron found in a small space between dendrites and axons is known as a synapse. The space of synapses is applicable for shipping and receiving all information processes from the senses. Any information entered will be encoded in the form of electrical signals. All electrical signals into the synapses are counted and calculated. When the number of electrical signals regardless of the limits or thresholds specified in the synapse, the synapses react to a new electrical signal input to be used by the next neuron. If the electrical signals cannot be separated from the predetermined threshold, then the synapses will be retarded. Retardation of synapses causes obstruction of the relationship between the two neurons.

In line with the biological neuron model, McCulloch and Pitt (1943) proposed a model neuron that has the characteristics of the transmission and receipt of information process that is similar to the process that occurs in biological neurons. This neuron modeling was becoming a reference in the development of ANN model at current state. A neuron plays a role in determining the function and operation of the network. The mathematical models of neurons, which are commonly used in the ANN model is shown in Fig. 2.2.

Neuron modeling based on Fig. 2.2 can be represented by the following mathematical equation:

$$u_{(k)} = \sum_{j=1}^{n} w_{kj} x_j \text{ and } y_{(k)} = \varphi\left(u_{(k)}\right) + b_{(k)} \tag{2.1}$$

where $u_{(k)}$ is the output of the adder function neuron model, x_j is data or input signal on path synapse j, and w_{kj} is the weighted in the path of synapse j to k neuron. The output of the neuron is represented by $y_{(k)}$, where it is dependent on the activation

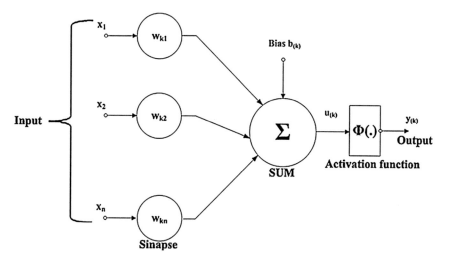

Fig. 2.2 Mathematical modeling of neuron

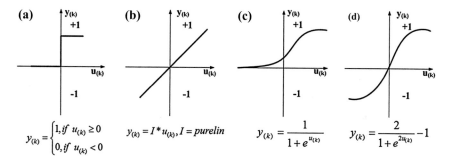

Fig. 2.3 Activation function for (**a**) the fixed restrictor, (**b**) purelin, (**c**) log-sigmoid, and (**d**) bipolar sigmoid

function $\varphi(\cdot)$ and the bias $b_{(k)}$. There are several types of activation functions that were used in modeling neurons, some of them are fixed limiter function, linear function, sigmoid function, and bipolar sigmoid function as shown in Fig. 2.3 (Duch and Jankowski 1999; Dorofki et al. 2012).

2.1.2 Architecture

Connections between neurons with other neurons will form a layer pattern, so-called net architecture. Normally, ANN architecture consists of three different layers. The first layer is called the input layer. This layer acts as a receiver of data or input from the external stimuli. Incoming data is then sent to the next layer. In this layer, the number of neurons can be more than one. There are no binding rules for determining the number of neurons; it depends on the number of entries to be used in the network. The next layer is a hidden layer. This layer contains neurons that can receive data or electrical signal than the previous layer of the input layer. Data or electrical signal that goes into these layers is processed using the functions available such as arithmetic, mathematics, etc. The hidden layer can contain one or more neurons, which depends on the suitability and complexity of the case at hand. Data processing results of this layer is then routed to the output layer. Output layer plays a role in determining the validity of data that are analyzed based on the existing limits in the activation function. The output of this layer can be used as a determinant of the outcome of the case at hand.

Based on the pattern of connections between neurons in the ANN, ANN architecture is divided into two types such as feedforward neural network and feedback neural network (Jain et al. 1996; Tang et al. 2007; Haykin 2009). Figure 2.4 shows the taxonomy of both the ANN architectures. Feedforward neural network is an ANN that does not have a feedback link on architecture. Data or incoming signals are allowed only to move in one direction only. This means that the output of each layer will not give any effect to the previous layer. In the architecture, it can be developed using a single layer or multiple layers. Usually, the multilayer component consists of three layers, namely a layer of input, output, and hidden. In a multilayer, hidden layer component plays a role in increasing the ability of computing power. One-layer perceptron, multilayer perceptron, and radial basis function are types of ANNs using feedforward neural networks.

Another architecture is a feedback neural network or repetitive. It has a design similar to the architecture of feedforward neural networks. However, in an architectural design there are additional feedbacks slow or feedback on the previous layer. This means the data or electrical signals that are allowed to propagate forward and feedback can be an input to the neurons before. This network is used for dynamic

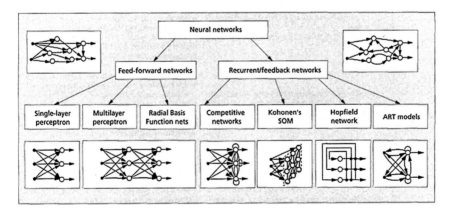

Fig. 2.4 Taxonomy of neural network architecture of feedforward and feedback neural networks adopted from Jain et al. (1996)

applications such as adaptive control. Hopfield networks, Elman network, and Jordan network are some examples of the types of ANNs using feedback neural network.

2.1.3   Learning Process

ANN learning algorithm plays a role in the process of modifying the parameters and the value in the network to adapt its environment. The use of learning algorithms allows ANN assembles themselves for giving consistent response to input into the network. During the learning process, the parameters and the weights of synapses that are in the network will be modified. This is a form of response to the input stimulus to the output produced in accordance with the desired output. Level of learning will expire when the resulting output was consistent with the desired output.

To understand or design a learning process on ANN, there are three steps that need to be done by the designer (Jain et al. 1996). The steps are (1) learning paradigm, which refers to a process where a designer in building a system needs to choose the learning process in accordance with the information environment of the system; (2) learning algorithm, which refers to a learning rule that is used to modify the parameters and weights of synapses in the ANN series; and (3) finally, it is important to assess how much the network can be learned (capacity) and how many samples are required for training (sample complexity) as well as how fast the system can learn (time complexity).

Refers to the type of learning in the ANN, two types of learning processes have been widely adopted, namely supervised and unsupervised learning. Apparent differences between both are on the information provided by the network. Usually, the information given to supervised learning is in the form of sample patterns that have been marked or labeled, while in the unsupervised learning it occurs oppositely. Thus, for unsupervised learning it worked at random.

In supervised learning, a pattern that was given to the network has been known its output. Each incoming signals into a single neuron will continue to spread out along the network until the end layer of neurons in the output layer. In the final layer, the output pattern will be generated and then compared with the desired output pattern. Upon the occurrence of an error signal during the process of comparison between the output patterns generated by the pattern of the desired output, then the process should be modified to adjust the network weights so that the actual output will be in accordance with the desired output.

In contrast to supervised learning, unsupervised learning does not have guidelines or target output in the learning process. The network only receives many samples of an input and then puts the sample in any way into some classes or category. When the stimulus was given to the input layer, the response in the form of production category or class will have similar characteristics to the input stimulus. In contrast, the network will form a new coding, which led to a new class or category (Haykin 2009).

2.2   Adaptive Neuro-Fuzzy Interference System

Modify network-based fuzzy inference (ANFIS) is a combination of two soft-computing methods of ANN and fuzzy logic (Jang 1993). Fuzzy logic has the ability to change the qualitative aspects of human knowledge and insights into the process of precise quantitative analysis. However, it does not have a defined method that can be used as a guide in the process of transformation and human thought into rule base fuzzy inference system (FIS), and it also takes quite a long time to adjust the membership functions (MFs) (Jang 1993). Unlike ANN, it has a higher capability in the learning process to adapt to its environment. Therefore, the ANN can be used to automatically adjust the MFs and reduce the rate of errors in the determination of rules in fuzzy logic. This section will describe in details of the architecture of ANFIS, FISs, and network flexibility, and hybrid learning algorithm.

2.2.1   Fuzzy Inference System

A FIS was built on the three main components, namely basic rules, where it consists of the selection of fuzzy logic rules "If-Then;" as a function of the fuzzy set membership; and reasoning fuzzy inference techniques from basic rules to get the output. Figure 2.5 shows the detailed structure of the FIS. FIS will work when the input that contains the actual value is converted into fuzzy values using the fuzzification process through its membership function, where the fuzzy value has a range between 0 and 1. The basic rules and databases are referred to as the knowledge base, where both are key elements in decision-making. Normally, the database contains definitions such as information on fuzzy sets parameter with a function that has been defined for every existing linguistic variable. The development of a database typically includes defining a universe, determination of the number of linguistic values to

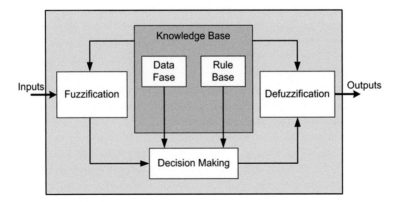

Fig. 2.5 Fuzzy inference system

be used for each linguistic variable, as well as establish a membership function. Based on the rules, it contains fuzzy logic operators and a conditional statement "If-Then." The basic rules can be constructed either from a human or automatic generation, where the searching rules using input–output data numerically. There are several types of FIS, namely Takagi–Sugeno, Mamdani, and Tsukamoto (Cheng et al. 2005). A FIS of Takagi–Sugeno model was found to be widely used in the application of ANFIS method.

2.2.2 *Adaptive Network*

Adaptive network is one example of feedforward neural network with multiple layers (see Fig. 2.6). In the learning process, these networks often use supervised learning algorithm. In addition, adaptive network has the architecture characteristics that consists of a number of adaptive nodes interconnected directly without any weight value between them. Each node in this network has different functions and tasks, and the output depends on the incoming signals and parameters that are available in the node. A learning rule that was used can affect the parameters in the node and it can reduce the occurrence of errors at the output of the adaptive network (Jang 1993).

In learning the basic adaptive network, it is normally using gradient descent or back propagation and the chain rule. All this learning algorithms had been proposed by Werbos in 1970 (Jang 1993). Till date, gradient descent or back propagation is still used as a learning algorithm in an adaptive network. Even so, there are still found weaknesses in the backpropagation algorithm and further can reduce the capacity and accuracy of adaptive networks in making decisions. The slow convergence rate and tend to always stuck in local minima are major problems on backpropagation algorithm. Therefore, Jang (1993) have proposed an alternative learning algorithm, namely hybrid learning algorithm, which has the better ability to accelerate convergence and avoid the occurrence of trapped in local minima.

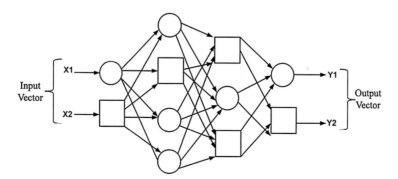

Fig. 2.6 Adaptive network

2.2.3   ANFIS Architecture

ANFIS architecture is an adaptive network that uses supervised learning on learning algorithm, which has a function similar to the model of Takagi–Sugeno fuzzy inference system. Figure 2.7a, b shows the scheme fuzzy reasoning mechanism for Takagi–Sugeno model and ANFIS architecture. For simplicity, assume that there are two inputs x and y, and one output f. Two rules were used in the method of "If-Then" for Takagi–Sugeno model, as follows:

$$\text{Rule } 1 = \text{If } x \text{ is } A_1 \text{ and } y \text{ is } B_1 \quad \text{Then } f_1 = p_1 x + q_1 x + r_1$$
$$\text{Rule } 2 = \text{If } x \text{ is } A_2 \text{ and } y \text{ is } B_2 \quad \text{Then } f_2 = p_2 y + q_2 y + r_2$$

where A_1, A_2 and B_1, B_2 are the membership functions of each input x and y (part of the premises), while p_1, q_1, r_1 and p_2, q_2, r_2 are linear parameters in part-Then (consequent part) of Takagi–Sugeno fuzzy inference model.

Referring to Fig. 2.7, ANFIS architecture has five layers. The first and fourth layers contain an adaptive node, while the other layers contain a fixed node. A brief description of each layer is as follows:

Layer 1: Every node in this layer adapts to a function parameter. The output from each node is a degree of membership value that is given by the input of the

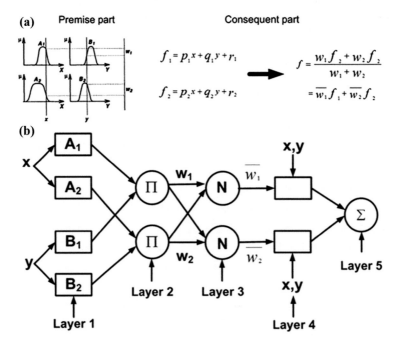

Fig. 2.7 a Sugeno fuzzy interference system "*If-Then*" and fuzzy logic mechanism. **b** ANFIS architecture (Suparta and Alhasa 2013)

membership functions. For example, the membership function can be a Gaussian membership function (Eq. 2.2), a generalized bell membership function (Eq. 2.3), or another type of membership function.

$$\mu_{Ai}(x) = \exp\left[-\left(\frac{x - c_i}{2a_i}\right)^2\right]$$

(2.2)

$$\mu_{Ai}(x) = \frac{1}{1 + \left|\frac{x-c_i}{a_i}\right|^{2b}}$$

(2.3)

$$O_{1,i} = \mu_{Ai}(x), \quad i = 1, 2$$

(2.4)

$$O_{1,i} = \mu_{Bi-2}(y), \quad i = 3, 4$$

(2.5)

where μ_{Ai} and μ_{Bi-2} are the degree of membership functions for the fuzzy sets A_i and B_i, respectively, and $\{a_i, b_i, c_i\}$ are the parameters of a membership function that can change the shape of the membership function. The parameters in this layer are typically referred to as the premise parameters.

Layer 2: Every node in this layer is fixed or nonadaptive, and the circle node is labeled as Π. The output node is the result of multiplying of signal coming into the node and delivered to the next node. Each node in this layer represents the firing strength for each rule. In the second layer, the T-norm operator with general performance, such as the AND, is applied to obtain the output

$$O_{2i} = w_i = \mu_{Ai}(x) * \mu_{Bi}(y), \quad i = 1, 2$$

(2.6)

where w_i is the output that represents the firing strength of each rule.

Layer 3: Every node in this layer is fixed or nonadaptive and the circle node is labeled as N. Each node is a calculation of the ratio between the i-th rules firing strength and the sum of all rules' firing strengths. This result is known as the normalized firing strength.

$$O_{3i} = \bar{w}_i = \frac{w_i}{\sum_i w_i}$$

(2.7)

Layer 4: Every node in this layer is an adaptive node to an output, with a node function defined as

$$O_{4i} = \bar{w}_i f_i = \bar{w}_i (p_i x + q_i y + r_i)$$

(2.8)

where $\bar{w}_i$ is the normalized firing strength from the previous layer (third layer) and $(p_i x + q_i y + r_i)$ is a parameter in the node. The parameters in this layer are referred to as consequent parameters.

Layer 5: The single node in this layer is a fixed or nonadaptive node that computes the overall output as the summation of all incoming signals from the previous node. In this layer, a circle node is labeled as $\sum$.

$$O_{5i} = \sum_i \bar{w}_i f_i = \frac{\sum_i w_i f_i}{\sum_i w_i} \tag{2.9}$$

2.2.4 Hybrid Learning Algorithm

In the ANFIS architecture, the first layer and the fourth layer contain the parameters that can be modified over time. In the first layer, it contains a nonlinear of the premises parameter while the fourth layer contains linear consequent parameters. To update both of these parameters required a learning method that can train both of these parameters and to adapt to its environment. A hybrid algorithm proposed by Jang (1993) will be used in this study to train of these parameters. The use of this algorithm is due to the backpropagation algorithm that was used to train the parameters that exist in the adaptive networks found problematic especially in a slow convergence rate and tend to be trapped in local minima.

There are two parts of a hybrid learning algorithm, namely the forward path and backward path. In the course of the forward path, the parameters of the premises in the first layer must be in a steady state. A recursive least square estimator (RLSE) method was applied to repair the consequent parameter in the fourth layer. As the consequent parameters are linear, then RSLE method can be applied to accelerate the convergence rate in hybrid learning process. Next, after the consequent parameters are obtained, input data is passed back to the adaptive network input, and the output generated will be compared with the actual output.

While backward path is run, the consequent parameters must be in a steady state. The error occurred during the comparison between the output generated with the actual output is propagated back to the first layer. At the same time, parameter premises in the first layer are updated using learning methods of gradient descent or back propagation. With the use of hybrid learning algorithm that combines RSLE and the gradient descent methods, it can ensure the convergence rate is faster because it can reduce the dimensional search space in the original method of backpropagation (Nayak et al. 2004). One level of hybrid learning is called epochs. Table 2.1 describes briefly a hybrid learning process in ANFIS.

Table 2.1 Hybrid learning process

Type	Path forwards	Path backwards
Premise parameter	Fixed	Gradient descent
Consequent parameter	RSLE	Fixed
Signal	Node output	Error rate

2.2.4.1 BackPropagation Learning for Parameter Premises

The premise parameters $\{a, b, c\}$ in Eqs. 2.3 and 2.4 are adaptive parameters that can be trained to get the parameters in accordance with its environments. Suppose to have an adaptive network and similar to the Fig. 2.7b, where the network consists of five layers and has a total of $N(L)$ node in layer-L, then the number of square error in the L layer to p data is $1 \leq p \leq N$, and it can be defined as follows (Jang 1993; Jang and Sun 1995):

$$E_p = \sum_{k=1}^{N(L)} d_k - X_{k,p}^L \qquad (2.10)$$

where d_k is the k-th component of the vector of the desired output, while $X_{k,p}^L$ is k-th component of the vector of actual output generated by adaptive network with input from the input vector p. The main goal of adaptive learning system is to reduce errors that occur in the Eq. 2.10.

An early stage of learning begins by calculating the error rate of the output i-th node and L layer, with derivation equation as follows:

$$\varepsilon_{L,i} = \frac{\partial E_p}{\partial_{i,p}^L} = -2\left(d_{i,p} - X_{i,p}^L\right) \qquad (2.11)$$

For internal nodes in the l layer at i position, the error rate can be calculated using the Chain Rule

$$\frac{\partial E_p}{\partial X_{l,i}} = \sum_{m=1}^{N(l+1)} \frac{\partial E_p}{\partial X_{m,p}^{l+1}} \frac{\partial X_{m,p}^{l+1}}{\partial X_{m,p}^{l+1}} \qquad (2.12)$$

with $0 \leq l \leq L - 1$. Internal node error signal can be expressed as a linear combination of the error rate in the layer node l $(l + 1)$. Equation 2.12 is used to calculate the error signal at i-th layer node to l $(l < L)$, while the use of Eq. 2.12 to reach the final layer. Further, when α is a parameter used in some node, and then the equation will be obtained as follows:

$$\frac{\partial E_p}{\partial \alpha} = \sum_{x^* \in S} \frac{\partial E_p}{\partial x^*} \frac{\partial x^*}{\partial \alpha} \qquad (2.13)$$

where S is the set of nodes containing the parameter α, so that the whole issue of measurement error of α will produce Eq. (2.14)

$$\frac{\partial E}{\partial \alpha} = \sum_{p=1}^{p} \frac{\partial E_p}{\partial \alpha} \qquad (2.14)$$

with steepest gradient descent method, the equation for repairing parameter α is obtained:

$$\Delta\alpha = -\eta\frac{\partial E}{\partial \alpha} \tag{2.15}$$

with η is the learning rate process and stated as follows:

$$\eta = \frac{k}{\sqrt{\sum \alpha\left(\frac{\partial E}{\partial \alpha}\right)^2}} \tag{2.16}$$

and k is the step size, which can be changed in order to accelerate the convergence rate in adaptive networks.

2.2.4.2   Learning to Parameter Consequent RSLE

During the premises parameter in a steady state, then all output derived from the consequent parameters can be specified in a combination linear equation (Jang 1993; Jang and Sun 1995):

$$\begin{aligned}
f &= \bar{w}_1 f_1 + \bar{w}_2 f_2 \\
&= \bar{w}_1(p_1 x + q_1 y + r_1) + \bar{w}_2(p_2 x + q_2 y + r_2) \\
&= (\bar{w}_1 x)p_1 + (\bar{w}_1 y)q_1 + (\bar{w}_1)r_1 + (\bar{w}_2 x)p_2 + (\bar{w}_2 y)q_2 + (\bar{w}_2)r_2
\end{aligned} \tag{2.17}$$

When N training data are given to Eq. 2.17, then the equation will be obtained as follows:

$$(\bar{w}_1 x)_1 p_1 + (\bar{w}_1 y)_1 q_1 + (\bar{w}_1)_1 r_1 + (\bar{w}_2 x)_2 p_2 + (\bar{w}_2 y)_2 q_2 + (\bar{w}_2)_2 r_2 = f1$$

$$\vdots \tag{2.18}$$

$$(\bar{w}_1 x)_n p_1 + (\bar{w}_1 y)_n q_1 + (\bar{w}_1)_n r_1 + (\bar{w}_2 x)_n p_2 + (\bar{w}_2 y)_n q_2 + (\bar{w}_2)_n r_2 = fn$$

To simplify, Eq. 2.18 can be expressed in matrix form as shown in Eq. 2.19:

$$A\theta = y \tag{2.19}$$

where θ is the vector $M \times 1$. M refers to the number of elements that are consequent parameter set. While A is the vector $P \times M$, where P is the number of N data training provided to the adaptive network and y is the output vector $P \times 1$ whose elements are N number of output data of an adaptive network. Normally, the amount of training data is larger than the number of consequent parameters, so the best solution for θ is minimizing the squared error $\|A\theta = y^2\|$. By the least squares estimator (LSE), the equation for θ is defined as

$$\theta^* = \left(A^T A\right)^{-1} A^T y \tag{2.20}$$

where A^T is the inverse of A and if not singular, $\left(A^T A\right)^{-1}$ is the pseudo-inverse of A. By using a recursive LSE method, then the Eq. 2.20 becomes

$$\left.\begin{array}{l} \theta_{i+1} = \theta_i + P_{i+1} a_{i+1}\left(y_{i+1}^T - a_{i+1}^T \theta_i\right) \\ P_{i+1} = P_i - \frac{P_i + a_{i+1} a_{i+1}^T P_i}{1 + a_{i+1}^T P_i a_{i+1}}, \quad i = 0, 1 \ldots, P-1 \end{array}\right\} \tag{2.21}$$

where a_i^T is a row vector of the matrix A in Eq. 2.19, y_i is i-th element of y. P_i sometimes called a covariance matrix and is defined by the following equation:

$$P_i = \left(A^T A\right)^{-1} \tag{2.22}$$

2.3 Linear Regression

In general, regression is a statistical method that can provide information about the patterns of relationships between two or more variables. In the regression method, it is identified two types of variables, namely (1) response variable or known also as the dependent variable; this variable is affected by other variables and usually denoted by Y, and (2) predictor variables are also known as independent variables, which are variables that are not affected by other variables and are usually denoted by X (Shafiullah et al. 2010).

The main goal in the regression analysis is to create a mathematical model that can be applied to forecast the values of the dependent variable based on the values of any variables. In use, the regression analysis is divided into two simple linear and multiple linear regressions. A simple regression analysis is a relationship between two variables, which are independent, and the dependent variables. In the multiple linear regression analysis, the relationship is found between three or more variables, which contain at least two independent variables and one dependent variable.

In the multiple linear regressions, the form of equation containing two or more variables is written as follows:

$$Y = \beta_0 + \beta_1 X_1 + \beta_2 X_2 + \beta_m X_m, \quad m = 1, 2, 3, \ldots, n \tag{2.23}$$

where β_0 is a cutoff and $\beta_1 \ldots \beta_m$ are the regression coefficients. To obtain the values of the intercept and the regression coefficient in Eq. 2.23, the least squares method is frequently used (Brown 2009). Further, the use of such methods will be described in detail in Chap. 4 to develop an estimation model for ZPD.

References

Brown SH (2009) Multiple linear regression analysis: a matrix approach with Matlab. Ala J Math 34:1–3

Cheng CT, Lin, JY, Sun YG, Chau K (2005) Long-term prediction of discharges in Manwan Hydropower using adaptive-network-based fuzzy inference systems models. Adv Nat Comput 1152-1161

Dorofki M, Elshafie AH, Jaafar O, Karim OA, Mastura S (2012) Comparison of artificial neural network transfer functions abilities to simulate extreme runoff data. 2012 International conference on environment, energy and biotechnology, pp 39–44

Duch W, Jankowski N (1999) Survey of neural transfer functions. Neural comput Surv 2:163–212

Haykin S (ed) (2009) Neural network and machine learning. Pearson Prentice Hall, New York

Jain AK, Jianchang M, Mohiuddin KM (1996) Artificial neural networks: a tutorial. Computer 29 (3):1–44

Jang JSR (1993) ANFIS: adaptive network-based fuzzy inference systems. IEEE Trans Sys Man Cybern 23:665–685

Jang JS, Sun CT (1995) Neuro-fuzzy modeling and control. Proc IEEE 83(3):378–406

McCulloch WS, Pitts WH (1943) A logical calculus of the ideas immanent in nervous activity. Bull Math Biophys 5:115–133

Nayak PC, Sudheer KP, Rangan DM, Ramasastri KS (2004) A neuro-fuzzy computing technique for modeling hydrological time series. J Hydrol 291(1):52–66

Shafiullah GM, Ali AS, Thompson A, Wolfs PJ (2010) Predicting vertical acceleration of railway wagons using regression algorithms. IEEE Trans Intel Syst 11(2):290–299

Suparta W, Alhasa KM (2013) A comparison of ANFIS and MLP models for the prediction of precipitable water vapor. 2013 IEEE international conference on space science and communication (IconSpace), pp 243–248

Tang H, Tan CK, Yi Z (2007) Neural networks: computational models and applications. Stud Com Intell, vol 53. Springer, Berlin

Chapter 3
Tropospheric Modeling from GPS

Abstract The dynamics of the neutral atmosphere is of great interest to a mete-orologist who predicts weather and climatologist who performs climate modeling. Modeling the effect of GPS signals for the above applications require information about the properties of the atmosphere. This chapter provides a modeling of tropospheric delay from the effect of the propagation GPS signals in the atmosphere. The modeling will include the overview of the empirical models of zenith tropospheric delay together with the mapping function.

Keywords Atmosphere · GPS signals · Refractive index · Tropospheric delay modeling · Mapping function

3.1 The Neutral Atmosphere and Its Composition

Monitoring of the dynamics of the atmosphere shows that they are composed of several chemically distinct gasses, the relative amounts of which within the lower atmosphere may be determined. The composition and structure of this unique resource are important keys to understanding circulation in the atmosphere, short-term local weather patterns and long-term global climate changes.

Characterizing the atmosphere, by the way, radio wave is propagated that leads to a subdivision of neutral atmosphere and ionosphere. The neutral atmosphere layer consists of three temperature-delineated regions: the troposphere, the stratosphere and part of the mesosphere. It is often simply referred to as the *troposphere* because in radio wave propagation, the troposphere effects dominate. Hence, to the GPS researcher, the "troposphere" is generally referred to the neutral atmosphere at altitudes 0–40 km (Gregorius and Blewitt 1999). On the other hand, when speaking of the troposphere, it will be clear from the context, whether it referred to the neutral atmosphere or the specific layer.

The layers of the troposphere are defined by their characteristics such as temperature, pressure, and chemical composition. Pressure and density decrease as a

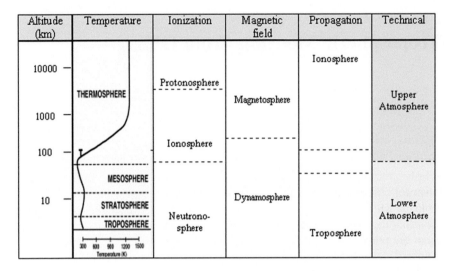

Altitude (km)	Temperature	Ionization	Magnetic field	Propagation	Technical

Fig. 3.1 Possible subdivision schemes of the Earth's atmosphere adapted from Seeber (1993)

function of altitude following the exponential barometric law. In general, the temperature in the troposphere decreases linearly with height at a rate of 6.5 °C km^{-1} (on average). The actual value of this temperature gradient is a function of height, season and geographical location. At the top of the *troposphere*, at a boundary layer between 12 and 18 km (mean sea level, MSL), the temperature remains approximately constant at a level of −60 to −80 °C. However, this boundary still has weather such as clouds formation and precipitation, wind blows, and the atmosphere interacts with the surface of the Earth below. This part of the neutral atmosphere is called the *tropopause*. The upper part above the tropopause is referred to as the *stratosphere*, up to an altitude of 40 km temperature increases again in the stratosphere up about 50 km altitude (as the *mesosphere*). The stratosphere is mainly responsible for absorbing the ultraviolet radiation. Between 50 and 80 km above MSL, the temperature drops again in the mesosphere. At the outer reaches of the Earth, atmosphere is the *thermosphere* with an initial slow temperature increase. Figure 3.1 shows the detailed subdivisions of the atmosphere with characteristic features such as temperature, ionization, and propagation (Seeber 1993).

3.2 Tropospheric Delay Modeling

Yuan et al. (1993) described that the troposphere affects the propagation of GPS radio signals in two ways. First, waves travel slower in the atmosphere ('bending effect') than they are in free-space. Second, they travel in a curved path rather than a straight-line ('geometrical delay or excess path delay'). Both effects arise significantly due to the refractivity variations in the atmosphere along the ray path and the modeling will take into account of these effects.

3.2.1 Refraction of GPS Signals in the Troposphere

Refraction effects are generally caused by an inhomogeneous propagation medium. The refractive index is often thought of as an "optical density" and for the ordinary ray it is constant and independent of direction. When the radio signals traverse the Earth's atmosphere, they are affected significantly by variations in the refractive index of the troposphere. Refraction bends the ray path and thereby lengthens it, further increasing the delay. The refractive index of a material is the factor by which the phase velocity of electromagnetic radiation is slowed in that material, relative to its velocity in a free-space. The tropospheric propagation delay can be determined from models and approximations of the atmospheric profiles.

The refractive index of the troposphere is constituent of gasses slightly greater than unity. The resulting decrease in velocity increases the time taken for signal to reach a receiver's antenna, thereby increasing the equivalent path length. The combination of these two effects is called the troposphere refraction component of propagation delay. Both L1 and L2 frequencies of GPS are affected by atmospheric refraction. This refractive delay obtained from biases between the satellite receiver range measurements.

3.2.1.1 Refractive Index

Like all electromagnetic waves, the ranging signals broadcasted by the GPS satellites can be described by Maxwell's equations. The propagation media in the equations are characterized by the magnetic permeability (μ) and the electric permittivity (ε). The velocity of an electromagnetic wave is characterized by the refractive index, n. These represent the ratio of the free-space speed (c) of electromagnetic wave to its media speed (v) and are related by Maxwell's equation (Brunner 1993). Therefore, the refractive index of a medium is given as

$$n = \frac{c}{v}\sqrt{\varepsilon \mu} \qquad (3.1)$$

Solution of Maxwell's equation can be difficult to obtain if μ and ε are functions of position. Basically, Snell's Law equation is commonly used to determine a refractive index for a simple case with two or three different mediums. However, one method based on the first principle of Newton's second law (see Griffith 1999) can be used to show where the neutral part of atmosphere acts as a nondispersive medium for the radio frequency.

3.2.1.2 Refractivity

The refractivity of the atmosphere determines the amount of "bending" of the radio waves. The refractive index of moist air is different from unity because its constituents suffer polarization induced by the electromagnetic field of the radio signals. As the electromagnetic waves in the atmosphere propagate just slightly slower than in a free-space, the refractive index is close to unity in the terrestrial atmosphere. It is convenient to define the refractivity (Brunner 1993):

$$N = 10^6(n - 1) \tag{3.2}$$

where n is the refractive index of radio wave in an air at ambient condition and N is the total refractivity of radio wave.

In the equation of state, total refractivity is a function of temperature, partial pressure of dry air, and partial pressure of water vapor that can be derived using the following expression (Smith and Weintraub 1953):

$$N = N_d + N_w = \underbrace{k_1 \frac{P_d}{T_K}}_{\text{dry}} + \underbrace{k_2 \frac{P_w}{T_K}}_{\text{dipole moment}} + \underbrace{k_3 \frac{P_w}{T_K^2}}_{\text{dipole orientation}} \tag{3.3}$$

where k_i ($i = 1\ldots3$) is the refraction constants are empirically determined and the most significant recent evaluations of the refractivity constants are summarized in Table 3.1, P_d is the partial pressure of dry air (mbar), T_K is the surface air temperature (Kelvin) and P_w is the partial pressure of water vapor (mbar).

Thayer (1994) took into account the nonideal gaseous behavior of the atmosphere and improved the refractivity formula as shown in Eq. 3.3. This reduced the computation uncertainty of 0.6 % before down to 0.02 %. Therefore, the refractivity N can be written as

$$N = k_1 \left(\frac{P_d}{T_K} \right) Z_d^{-1} + \left(k_2 \frac{P_w}{T_K} + k_3 \frac{P_w}{T_K^2} \right) Z_w^{-1} \tag{3.4}$$

Table 3.1 Determinations of the refractivity constants (Bevis et al. 1994; Suparta 2008)

Reference	k_1 (K mbar^{-1})	k_2 (K mbar^{-1})	k_3 (K^2 mbar^{-1}) × 10^5
Smith and Weintraub (1953)	77.61 ± 0.01	72 ± 9	3.75 ± 0.03
Boudouris (1963)	77.59 ± 0.08	72 ± 11	3.75 ± 0.03
Thayer (1974)	77.61 ± 0.01	47.79 ± 0.08	3.776 ± 0.04
Hill et al. (1982)	–	98 ± 1	3.583 ± 0.03
Hill (1988)	–	102 ± 1	3.578 ± 0.03
Clynch (1990)	77.604 ± 0.02	75 ± 0.1	3.75 ± 0.01
Bevis et al. (1992)	77.60 ± 0.05	70.4 ± 2.2	3.739 ± 0.012
Bevis et al. (1994)	77.60 ± 0.09	69.4 ± 2.2	3.701 ± 1200

In the above equation, P_w is obtained from relative humidity (H) as recommended by World Meteorological Organization Technical Note No. 8 (WMO 2000) and given by

$$P_w = \frac{H}{100} \exp\left(-37.2465 + 0.213166\, T_K - 2.56908 \times 10^{-4}\, T_K^2\right) \quad (3.5)$$

Both the dimensionless Z_d^{-1} and Z_w^{-1} are the inverse compressibility factors for dry air and water vapor constituents, respectively, to account for nonideal gas behavior. They have been experimentally determined by Owens (1967) and given as follows

$$Z_d^{-1} = 1 + (P - P_w)\left[57.97 \times 10^{-8}\left(1 + \frac{0.52}{T_K}\right) - 9.4611 \times 10^{-4}\frac{T}{T_K^2}\right] \quad (3.6)$$

$$Z_w^{-1} = 1 + 1650\frac{P_w}{T_K^3}\left(1 - 0.01317\, T + 1.75 \times 10^{-4}\, T^2 + 1.44 \times 10^{-6}\, T^3\right) \quad (3.7)$$

The first term of the Thayer Eq. 3.4 can be reformulated as a function of total moist air density (ρ_{tot}), allowing its direct integration by applying the hydrostatic equation (Davis et al. 1985). Consequently, the refractivity constant k_2 is also substituted with a new constant k_2' (Bevis et al. 1994) and the final expression for the total refractivity can be given as a sum of a hydrostatic (as opposed to dry) and wet components. The expression for total refractivity from Eq. 3.4 can be rewritten by separating the dry and wet terms as follows:

$$N = \underbrace{k_1\left(\frac{P_d}{T_K}\right)Z_d^{-1}}_{\text{dry}} + \underbrace{\left(k_2\frac{P_w}{T_K} + k_3\frac{P_w}{T_K^2}\right)Z_w^{-1}}_{\text{wet}} \quad (3.8)$$

By introducing ρ_{tot} (Wallace and Hobbs 1997) and measured quantity of pressure, P is, respectively, given as

$$\rho_{tot} = \rho_d + \rho_w \quad \text{and} \quad P = P_d + P_w \quad (3.9)$$

The first ideal gas equation, applied to dry air (ρ_d) and water vapor (ρ_w) were introduced by Spilker (1996), respectively.

$$P_d = \rho_d R_d T_K Z_d \quad \text{and} \quad P_w = \rho_w R_w T_K Z_w \quad (3.10)$$

A relation between molar mass of dry air and water vapor, and universal gas constant in the equation of state for ideal gasses in Eq. 3.10 can be approximated by

$$R_d = \frac{R}{M_d}, \quad R_w = \frac{R}{M_w}, \quad \text{and} \quad \frac{R_d}{R_w} = \frac{M_w}{M_d} \quad (3.11)$$

Let us now consider the dry part of the refractivity formula in Eq. 3.8

$$k_1 \left(\frac{P_d}{T_K}\right) Z_d^{-1} = k_1 \frac{P_d}{T_K} \frac{\rho_d R_d T_K}{P_d} = k_1 (\rho_{tot} - \rho_w) R_d = k_1 \rho_{tot} R_d - k_1 \rho_w R_d \quad (3.12)$$

Considering the ideal gas equation for water vapor in the Eqs. 3.10 and 3.11, Eq. 3.12 can be written as

$$k_1 \left(\frac{P_d}{T_K}\right) Z_d^{-1} = k_1 \rho_{tot} R_d - k_1 \frac{P_w}{T_K} \frac{M_w}{M_d} Z_w^{-1} \quad (3.13)$$

Substitution of this expression into the total refractivity formula in Eq. 3.8 yields

$$N = k_1 R_d \rho_{tot} + \left(k_2 \frac{P_w}{T_K} - k_1 \frac{P_w}{T_K} \frac{M_w}{M_d} + k_3 \frac{P_w}{T_K^2}\right) Z_w^{-1} \quad (3.14)$$

where the dry inverse compressibility factor is eliminated. The total refractivity is then given as

$$N = k_1 R_d \rho_{tot} + \left(k_2' \frac{P_w}{T_K} + k_3 \frac{P_w}{T_K^2}\right) Z_w^{-1} \quad (3.15)$$

with

$$k_2' = k_2 - k_1 \frac{M_w}{M_d} = (22.1 \pm 2.2) \quad (3.16)$$

where R is the universal gas constant (8314.34 J kmol^{-1} K^{-1}), R_d is the specific gas constant for dry air (287.054 J kg^{-1} K^{-1}), R_w is the specific gas constant for water vapor (461.5184 J mol^{-1} K^{-1}), ρ_{tot} is the total mass density (moist air density) of the troposphere (kg m^{-3}), ρ_d is the density of dry air (kg m^{-3}), ρ_w is the density of water vapor (kg m^{-3}), M_w is the molar mass of water vapor (28.9644 kg kmol^{-1}), and M_d is the molar mass of dry air (18.0152 kg kmol^{-1}). Expanding on development of refractivity as a function of wet and hydrostatic components, it is possible to examine their individual contribution to the tropospheric path delay.

3.2.2 Tropospheric Path Delay

There are two main parameters which play an important role during the propagation between transmitter (GPS) and receiver: pseudorange and carrier phases. All these propagation effects and time offsets have to be determined to account accurate estimation of position from range data. Thus, to understand comprehensively about tropospheric path delay modeling, the basic properties of radio waves propagation in the troposphere will first be described.

The basic physical law for the propagation is Fermat's principle: Light (or any electromagnetic wave) will follow the path between two points (P_1 and P_2 are ends of S) involving the least travel time. We define the electromagnetic (or optical) distance between source and receiver as

$$L = \int c\,dt = \int \frac{c}{v}\,dS = \int_S n(s)\,dS \tag{3.17}$$

where L is the delay of radio wave (so-called optical path length or electromagnetic distance, total tropospheric delay), $n(s)$ is the index of refraction which varies as a function position along the curved ray path L, S is the electromagnetic path, dS is infinitesimal parts of the path length, and c and v are speed of the radio signals in free-space and in medium, respectively. If we denote the geometrical distance or the straight-line (rectilinear) path in a free-space by

$$G = \int_G dG \tag{3.18}$$

where G is the geometrical distance and dG is an infinitesimal part of the path length in free-space.

Figure 3.2 shows the GPS signals traveling through the troposphere. The total delay, then, is the sum of these two components and can be written as

$$\Delta L = \int_S n(s)\,dS - G \tag{3.19}$$

or,

$$\Delta L = \int_S (n(s) - 1)dS - \int_G dG \tag{3.20}$$

$$\Delta L = \underbrace{\int_S [n(s) - 1]dS}_{\text{The slowing effect}} + \underbrace{[S - G]}_{\text{The bending effect}} \tag{3.21}$$

where ΔL is the total tropospheric delay stated in terms of equivalent increase in path length, S is the true path along L which the radio wave propagates and G is the

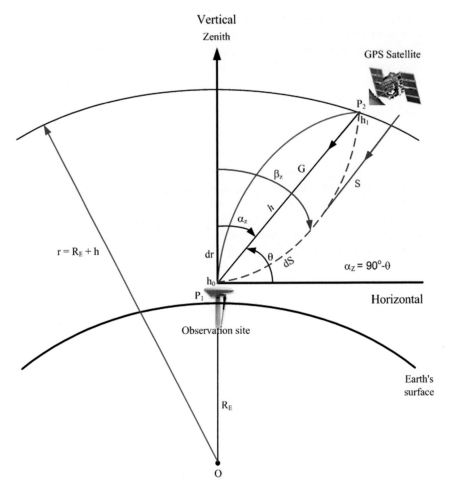

Fig. 3.2 GPS signals traveling through troposphere and the tropospheric path delay geometry (Suparta 2008)

shortest geometric path along which the signal would transverse, and assuming as $n = 1$. In the first term of Eq. 3.21, the integral is performed along the line increment dS of straight ray path (excess path delay) from a receiver to a GPS satellite.

The second term indicates the geometric delay due to ray path bending. The bending term $[S − G]$ is much smaller, about 1 cm or less, for path with elevation angle greater than 10°. Bending is about 1 mrad for a 15° elevation angle and its associated excess path length is about 1 cm, which is usually neglected since it represents ~ 0.1 % of the total path delay (see Bock and Doerflinger 2001). For rays oriented along the zenith and in the absence of horizontal gradients in index refractivity n, the ray path is straight-line and the bending term vanishes. Excess path length due to signal retarding in the troposphere in Eq. 3.21, or slant path delay (Davis 1985), is expressed as

$$\Delta L^z = \int_S [n(s) - 1]dS \qquad (3.22)$$

where ΔL^z is the total tropospheric delay in the zenith direction, which is referred as the ZTD.

However, in the troposphere several simplifying assumptions can be made to simplify Eq. 3.22. First, the atmosphere is assumed to be spherically symmetric, that is, the Earth is a sphere and the properties of the atmosphere vary only with geometric radius. In this way, the atmosphere can be considered layered with a refractive index characterizing each layer. Second, the atmosphere is usually assumed to be azimuthally symmetric, that is, with no variation of the refractive index in azimuth of each layer. In this way, the electromagnetic ray is confined to a plane defined by the start and end points of the ray and the geocentric. These assumptions allow us to represent the refractive index profile as a function of geocentric radial distance only, $n(h)$. On application of the refractivity in Eq. 3.2, Eq. 3.21 for S can then be written as

$$G + \Delta L^z = \int_{h_0}^{h_1} n(h) \sec \beta_z(h)dh \qquad (3.23)$$

where the refractive index is integrated along the path between points h_0 and h_1, which are the geocentric distance of the user's antenna and the geocentric distance of the 'top' of troposphere, respectively. Angle α_z is the true (unrefracted) satellite zenith angle and hence constant along the unrefracted path. Angle β_z is the actual (refracted) zenith angle of the ray path at distance h (see Fig. 3.2). The path delay is caused by variation of n from unity, hence

$$G + \Delta L^z = \int_{h_0}^{h_1} [n(h) - 1] \sec \beta_z(h)dh + \int_{h_0}^{h_1} \sec \beta_z(h)dh \qquad (3.24)$$

This gives in the first term the excess delay equivalent path length and in the second term the geometric length along the curved path. To obtain the total tropospheric delay (ΔL^z), we can subtract the geometric distance in free-space to get the following integral equation (Langley 1996):

$$\Delta L^z = \int_{h_0}^{h_1} \underbrace{[n(h) - 1] \sec \beta_z(h)dh}_{symmetric} + \left[\underbrace{\int_{h_0}^{h_1} \sec \beta_z(h)dh - \int_{h_0}^{h_1} \sec \alpha_z dh}_{\delta-asymmtric} \right] \quad (3.25)$$

To summarize, the first integral accounts for the difference between the electromagnetic distance and geometric distance along the ray path and the bracketed integrals account for curvature of the ray path, i.e., the difference between the refracted and rectilinear geometric distances. The total tropospheric delay by inserting Eq. 3.2, can be simplified as

$$\Delta L^z = 10^{-6} \int_{actual} N(h)dh + \delta \quad (3.26)$$

For easy modeling of the tropospheric delay, the total refractivity at distance h, $N(h)$ in the troposphere can be explicitly written as the contribution of a wet (N_w) and a hydrostatic (N_h) component. Therefore, Eq. 3.26 can be written as

$$\Delta L^z = \left(\int N_h dh + \int N_w dh \right) 10^{-6} + \delta \quad (3.27)$$

and symbolically, as

$$\Delta L^z = (L_h + L_w) + \delta \quad (3.28)$$

where L_h represents the hydrostatic delay and L_w is the wet delay.

Propagation delays at arbitrary elevation angles are determined from the zenith delay and arecalled the "mapping functions." As the zenith delay can be expressed as the sum of the hydrostatic and wet components, mapping functions can be developed in order to map separately the hydrostatic and wet components. Tropospheric delays increase with decreasing satellite elevation angle. This is accounted for by multiplying the zenith delay by a correction factor, m. In general, total tropospheric delay from Eq. 3.28, following Davis et al. (1985), can be rewritten as

$$\Delta L^z = (m_h ZHD + m_w ZWD) + \delta \quad (3.29)$$

where ΔL^z is the total delay along the zenith path called zenith path delay (ZPD), sometimes called the zenith total delay (ZTD) or zenith tropospheric delay (m). ZHD and ZWD are the hydrostatic zenith delay and the wet zenith delay, which both in meter, and θ is the satellite elevation angle (degrees). The last symbol in Eq. 3.29, δ is the tropospheric correction (recently, known as a gradient tropospheric delay, $\delta = m(\theta)\cot\theta\left[G_N\cos\alpha_z + G_E\sin\alpha_z\right]$ symmetric effects into account ($\delta = 0$, the asymmetric components are neglected by setting the cutoff elevation angle $\geq 10°$), and m is the obliquity factor from sec α_z ($\alpha_z = 90° - \theta$, is the azimuth angle), separated into m_h and m_w, the hydrostatic and wet mapping functions, respectively. G_N and G_E are the components of the gradient vector in the north and east directions, respectively.

3.3 Empirical Models of Tropospheric Delay

In the past several decades, a number of tropospheric propagation models have been reported in the scientific literature. Much research has gone into the creation and testing of tropospheric refraction models to compute the refractivity along the path of signal travel. The various tropospheric models differ primarily with respect to the assumption made regarding the vertical refractivity profiles and the mapping functions to map the delays to the arbitrary elevation angles. To model the tropospheric delay, many models use information about the surface pressure, temperature, and relative humidity to derive zenith or slant delay estimates. However, most models require certain conditions in, or make assumptions about, the atmosphere above the station. Among the commonly used models for the tropospheric delay are Saastamoinen (1972), Hopfield (1969), Modified Hopfield (Goad and Goodman 1974), Davis (1985), Herring (1992), Lanyi (1984), and Niell (1996, 2000). In this section, only the first three models are discussed. These models are most widely used due to their high accuracy, practicality, and suitable with the GPS measurements.

3.3.1 The Saastamoinen Model

The Saastamoinen model (SAAS) was developed for high elevation angles. This model has become popular among GPS users due to its accuracy. This model assumes that the atmosphere is in hydrostatic equilibrium, which follows from the ideal gas law. Under hydrostatic equilibrium, the local pressure, which is assumed isotropic, provides the balancing force against the atmospheric weight per unit area. Models for the ZTD, ZHD, and ZWD as derived by Saastamoinen (1972) will be described in this subsection.

Considering only the delay in the zenith direction, Eq. 3.25 reduces to

$$\Delta L^z = \text{ZTD} = \int_{h_0}^{h_1}[n(h) - 1]dh = 10^{-6}\int_{h_0}^{h_1}N(h)dh \qquad (3.30)$$

or, explicitly using Eq. 3.16,

$$\text{ZTD} = 10^{-6} \underbrace{\int_{h_0}^{h_1} k_1 R_d \rho_{\text{tot}} dh}_{\text{hydrostatic}} + 10^{-6} \underbrace{\int_{h_0}^{h_1} \left(\frac{k_2'}{T_K} + \frac{k_3}{T_K^2} \right) P_w Z_w^{-1} dh}_{\text{wet}} \qquad (3.31)$$

The first term in Eq. 3.31 represents the ZHD. By assuming a radio signal arrives from a zenith direction, the ZHD can be written as

$$\text{ZHD} = 10^{-6} k_1 R_d \int_{h_0}^{\infty} \rho_{\text{tot}}(h) dh \qquad (3.32)$$

Under the condition of hydrostatic equilibrium, the hydrostatic equations is

$$dP = -g(h)\rho_{\text{tot}}(h) dh \qquad (3.33)$$

where dP is the differential change in surface pressure (mbar), $g(h)$ is the acceleration due to gravity as a function of height (ms^{-2}), $\rho_{\text{tot}}(h)$ is the density of moist air as a function of height, and dh is the differential change in height (m).

Integrating Eq. 3.33 yields

$$\int_{P}^{0} dP = -\int_{h_0}^{\infty} \rho_{\text{tot}}(h) g(h) dh = -P \qquad (3.34)$$

Introducing the weighted mean gravity acceleration, g_m, the ZHD can be written as

$$\text{ZHD} = 10^{-6} k_1 R_d \frac{P}{g_m} \qquad (3.35)$$

The second term in Eq. 3.31 is ZWD. The ZWD can also be integrated after specifying suitable relationships for temperature and water vapor pressure with height. Unfortunately, water vapor is rarely in hydrostatic equilibrium and varies significantly throughout the troposphere; hence, specifying an accurate relationship with height is difficult. However, it is common in meteorology to model the average decrease of water vapor (or total pressure) with height as an exponential function with exponent γ. From Smith (1966), the mixing ratio (w) of water vapor to moist air has been given approximately by

$$w = w_0 \left(\frac{P}{P_0}\right)^{\gamma} \tag{3.36}$$

where the zero subscript indicate surface (i.e., MSL) value. However, $w = (M_w/M)(e_s/P)$, $w_0 = (M_w/M)(e_0/P_0)$ and by substitution and re-arrangement, we can obtain:

$$e_s = e_0 \left(\frac{P}{P_0}\right)^{\gamma+1} \tag{3.37}$$

This provides a relationship for the average decrease in water vapor pressure with height. Separating the two ZWD components in Eq. 3.32 and ignoring the wet compressibility factors we have

$$\text{ZWD} = 10^{-6} k_2' \int_{r_0}^{r_1} \frac{P_w}{T_K} dr + 10^{-6} k_3 \int_{r_0}^{r_1} \frac{P_w}{T_K^2} dr \tag{3.38}$$

By specifying a linear lapse rate (positive), β for temperature, the temperature throughout the troposphere can be represented as

$$T = T_0 \left(\frac{P}{P_0}\right)^{\frac{R_d \beta}{g}} \tag{3.39}$$

Combined with Eq. 3.35, allows for integration of Eq. 3.38. The formulation given by Askne and Nordius (1987) for ZWD is

$$\text{ZWD} = \frac{10^{-6} k_2' R_d}{g_m(\gamma+1)} P_w + \frac{10^{-6} k_3 R_d}{g_m(\gamma+1-\beta R_d/g_m)} \frac{P_w}{T_K} \tag{3.40}$$

where the mean temperature of the water vapor $T_m = T\left(1 - \frac{\beta R_d}{g_m(\gamma+1)}\right)$ units of Kelvin. By using the models in Eqs. 3.35 and 3.41, a general formulation for the ZTD is found as

$$\text{ZTD} = 10^{-6} k_1 \frac{R_d}{g_m} P + 10^{-6} \frac{R_d}{g_m} \left(\frac{k_2'}{(\gamma+1)} + \frac{k_3}{(\gamma+1-\beta R_d/g_m)T_K}\right) P_w \tag{3.41}$$

The ZTD model from Eq. 3.41 assumes that the delay caused by the ray bending and horizontal layer atmospheres is neglected. In general, because of a radio signal can come from slant directions, Saastamoinen (1972) and Hopfield (1969) develop a ZTD model by including slant delays (or slant tropospheric delay, STD) and internally cover a mapping function.

To derive of the ZTD model from Saastamoinen, we start with a truncated Taylor expansion of sec z in Fig. 3.3:

$$\sec z = \sec z_0 + \sec z_0 \tan z_0 \Delta z = \sec z_0(1 + \tan z_0 \Delta z) \qquad (3.42)$$

where, $\Delta z = z - z_0 = -\theta$ and $\tan z = r_0\, \theta/(r - R_{\mathrm{E}})$. So by approximating tan $z \approx \tan z_0$, Eq. 3.42 becomes

$$\sec z = \sec z_0 \left(1 - \tan^2 z_0 \frac{r - R_{\mathrm{E}}}{R_{\mathrm{E}}} \right) \qquad (3.43)$$

with this expression, the STD reads

$$\mathrm{STD} = 10^{-6} \int_{R_E}^{\infty} N \sec z \, \mathrm{d}r \qquad (3.44)$$

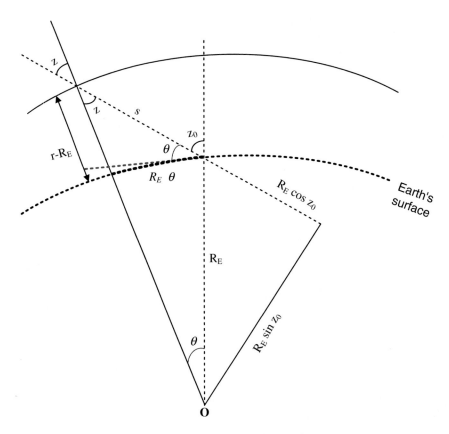

Fig. 3.3 Geometry of a ray arriving through a spherical atmosphere

$$= 10^{-6} \sec z_0 \left[\int_{R_E}^{\infty} N \, dr - R_E^{-1} \tan^2 z_0 \int_{R_E}^{\infty} N(r - R_E) dr \right] \tag{3.45}$$

The first term between the brackets in Eq. 3.45 is the zenith delay. The second term is a correction term of which the integral part can be subdivided into three sub-integrals

$$\int_{R_E}^{\infty} N \cdot (r - R_E) dr = \underbrace{\int_{R_E}^{r_T} N \cdot (r - R_E) dr}_{1} + \underbrace{\int_{r_T}^{\infty} N \cdot (r - r_T) dr}_{2} + \underbrace{(r_T - R_E) \int_{r_T}^{\infty} N \, dr}_{3} \tag{3.46}$$

where r_T is the radius of the tropopause, S is the traveled distance through the atmosphere, R_E is the radius of the Earth, z is the zenith angle at the top of the atmosphere, and z_0 is the zenith angle at the surface. Saastamoinen assumed the neutral atmosphere to consist of only two layers: the troposphere and the stratosphere. In this model, the troposphere is a polytrophic layer reaching up to r_T and the stratosphere is an isothermal layer, which for practical integration can be considered infinitely high. Each of the three integrals can be evaluated based on the refractivity profiles associated with the temperature profile.

The following evaluate of the integrals of Eq. 3.46, where Saastamoinen obtained his zenith delay model. In the troposphere, the temperature decreases with altitude. From this, the derivation of a pressure profile based on dry air we have differential equation

$$\frac{dP}{P_d} = -\frac{g_m}{R_d T} dh \tag{3.47}$$

The gravitation to be constant with height and equal to a mean value is considered.

$$g_m = \frac{\int_{h_0}^{\infty} \rho_m(h) g(h) dh}{\int_{h_0}^{\infty} \rho_m(h) dh} \tag{3.48}$$

For isothermal layer like the tropopause, the pressure profile is found by integration of Eq. 3.47

$$P_d = P_{d0} \exp\left(-\frac{h - h_0}{H}\right); \quad H = \frac{R_d T}{g_m} \tag{3.49}$$

In case of polytrophic layers, like the troposphere and stratosphere, the temperature lapse rate ($\beta = -\mathrm{d}T/\mathrm{d}H$) is assumed linear with height (H). We integrate the right-hand side of Eq. 3.48 over $\mathrm{d}T$,

$$P_d = P_{d0} \exp\left(-\frac{T}{T_0}\right)^{\mu+1}; \quad \mu = \frac{g_m}{R_d \beta} - 1 \tag{3.50}$$

From Eqs. 3.12, 3.49, and 3.50, the refractivity profile of dry air can also be derived. For polytrophic layers

$$\frac{N_d}{N_{d0}} = \frac{k_1 P_d / T}{k_1 P_{d0} / T_0} = \left(\frac{T}{T_0}\right)^{\mu}; \quad \mu = -1 \tag{3.51}$$

In an isothermal layer ($T = T_0$) we find

$$\frac{N_d}{N_{d0}} = \frac{P_d}{P_{d0}} = \exp\left(-\frac{h - h_0}{H}\right) \tag{3.52}$$

where P_{d0} is the pressure of dry air at the surface of the layer (mbar), N_{d0} is the dry refractivity at the surface of the layer, T_0 is the temperature at the surface of the layer (K), h_0 is the height above MSL at the surface of the layer (km), h is the height above MSL (km), and H is the scale height in km ($H = r - R_E$, see Fig. 3.3).

For the first integral in Eq. 3.48

$$N = N_0 \left(\frac{T}{T_0}\right)^{\mu} \quad \text{and} \quad T = T_0 - \beta H \Rightarrow \begin{cases} H = -\left(\frac{T_0}{\beta}\right)\left(\frac{T}{T_0} - 1\right); \\ \dfrac{\mathrm{d}H}{\mathrm{d}T} = -\dfrac{1}{\beta} \end{cases} \tag{3.53}$$

After some straightforward computations, this results in

$$\begin{aligned}
\int_{R_E}^{\infty} N \cdot (r - R_E)\mathrm{d}r &= \int_{T_0}^{T_T} N_0 \left(\frac{T}{T_0}\right)^{\mu}\left(-\frac{T}{\beta}\right)\left(\frac{T}{T_0} - 1\right)\frac{1}{\beta}\mathrm{d}T \\
&= \frac{R_d}{g_m^2(1 - R_d\beta/g_m)}\left[N_0 T_0^2 - N_T T_T^2\right] \\
&\quad - \frac{R_d}{g_m}(r_T - R_E)N_T T_T
\end{aligned} \tag{3.54}$$

where N_0 is the refractivity at Earth's surface (the index T stands for values at the tropopause), T_0 is the temperature at surface (or antenna) height (°C), β is the temperature lapse rate (0.0062 K km^{-1}), μ_m is the constant, $\mu_m = \frac{g_m}{R_d \beta} - 1$, and g_m is the mean acceleration gravity (9.784 ms^{-2}).

The second integral, can be evaluated by using the assumption of an exponential profile of Eq. 3.52 and effective height (H_m)

$$N = N_T \exp\left(-\frac{r - r_T}{H}\right); \quad H_m = \frac{R_d T_T}{g_m} \tag{3.55}$$

and results in

$$\int_{r_T}^{\infty} N(r - r_T)\,dr = \left(\frac{R_d}{g_m}\right)^2 N_T T_T^2 \tag{3.56}$$

The third integral is similar to Eq. 3.46 and results in

$$(r_T - R_E) \int_{r_T}^{\infty} N\,dr = (r_T - R_E)10^{-6} k_1 \frac{R_d}{g_m} P_T = \frac{R_d}{g_m}(r_T - R_E)N_T T_T \tag{3.57}$$

Summation of the three integral gives the total integral

$$\int_{R_E}^{\infty} N\,(r - r_T)\,dr = \left(\frac{R_d}{g_m}\right)^2 \left[\frac{N_0 T_0^2 - \left(R_d \beta/g_m^2\right) N_T T_T^2}{1 - R_d \beta/g_m}\right] \tag{3.58}$$

With Eq. 3.31, the total Saastamoinen model then becomes

$$\text{ZTD}_{\text{SAAS}} = 10^{-6} \sec z_0 \left[\int_{R_E}^{\infty} N\,dr - R_E^{-1} \tan^2 z_0 \cdot \int_{R_E}^{\infty} N(r - R_E)\,dr\right] \Rightarrow$$

$$\text{ZTD}_{\text{SAAS}} = 10^{-6} k_1 \frac{R_d}{g_m} \sec z_0 \left[P + \left(\frac{1255}{T_K} + 0.05\right) e_s - B(r) \tan^2 z_0\right] \tag{3.59}$$

where

$$B(r) = \frac{1}{R_E} \frac{g_m}{R_d} \frac{1}{k_1} \int_{r_0}^{\infty} N(r - R_E)\,dr \tag{3.60}$$

Tabular values for the correction term $B(r)$ are given by Saastamoinen (1972). The correction terms δR and $B(r)$ can be interpolated from Table 3.2. Saastamoinen did not mention the exact theoretical standard atmosphere he used to find the tabular values of $B(r)$. However, the standard values at MSL as also later used in the 1976 US Standard Atmosphere ($T_{\text{MSL}} = 288.15$ K, $P_{\text{MSL}} = 1013.25$ mbar), as well as the

Table 3.2 Correction terms for Saastamoinen neutral delay model (Hofmann-Wellenhof et al. 2001)

Zenith distance	Station height above sea level (km)							
	0	0.5	1	1.5	2	3	4	5
60°00′	0.003	0.003	0.002	0.002	0.002	0.002	0.001	0.001
66°00′	0.006	0.006	0.005	0.005	0.004	0.003	0.003	0.002
70°00′	0.012	0.011	0.010	0.009	0.008	0.006	0.005	0.004
73°00′	0.020	0.018	0.017	0.015	0.013	0.011	0.009	0.007
75°00′	0.031	0.028	0.025	0.023	0.021	0.017	0.014	0.011
76°00′	0.039	0.035	0.032	0.029	0.026	0.021	0.017	0.014
δR, m: 70°00′	0.050	0.045	0.041	0.037	0.033	0.027	0.022	0.018
78°00′	0.065	0.059	0.054	0.049	0.044	0.036	0.030	0.024
78°30′	0.075	0.068	0.062	0.056	0.051	0.042	0.034	0.028
79°00′	0.087	0.079	0.072	0.065	0.059	0.049	0.040	0.033
79°30′	0.102	0.093	0.085	0.077	0.070	0.058	0.047	0.039
79°45′	0.111	0.101	0.092	0.083	0.076	0.063	0.052	0.043
80°00′	0.121	0.110	0.100	0.091	0.083	0.068	0.056	0.047
B(r), mbar	0.156	1.079	1.006	0.938	0.874	0.757	0.654	0.563

values $r_0 = 6360$ km and $h_T = 15$ km, fit quite well. A slightly different table for the $B(r)$ values can be found in Saastamoinen (1972).

The ZTD of Saastamoinen model has refined his model by adding an additional term is used to account for the delay caused by the ray bending, δR as a complete form:

$$\mathrm{ZTD}_{\mathrm{SAAS}} = 10^{-6} k_1 \frac{R_{\mathrm{d}}}{g_{\mathrm{m}}} \sec z_0 \left[P + \left(\frac{1255}{T_{\mathrm{K}}} + 0.05 \right) P_{\mathrm{w}} - B(r) \tan^2 z_0 \right] + \delta R$$

(3.61)

where k_1 is the hydrostatic refractivity constant ($k_1 = 77.6 \pm 0.05$ K mbar^{-1}), $B(r)$ is the correction term of height dependent (mbar), δR is the correction term of ray bending (m), and z_0 is the zenith distance of the satellite or apparent zenith angle $z_0 = 90° - \theta$.

Looked at the first term in Eq. 3.61, the ZHD is with a mapping function $\sec z_0$. The mean gravitational acceleration depends on latitude and height of the antenna. Based on Saastamoinen (1972) approximation, the weighted mean gravity (g_{m}) is used to correct the gravitational acceleration at the center of mass of the vertical atmospheric column directly above the station depends on height at site and geodetic latitude, and is given as follows (Davis et al. 1985):

$$g_{\mathrm{m}} = 9.784 \, f(\varphi, h)$$

(3.62)

Finally, the expression for ZHD from Saastamoinen can be written as follows:

$$\text{ZHD}_{\text{SAAS}}(P, \varphi, h) = (2.2768 \pm 0.0024)\frac{P}{f(\varphi, h)} \qquad (3.63)$$

where $f(\varphi, h) = 1 - 0.00266 \cos(2\varphi) - 0.00028h$, is the correction factor for the local gravitational acceleration, φ is the site latitude (in degrees) and h is the height of the site above the ellipsoid (in km). Accurately, with Eq. 3.63, for any location on Earth, when the surface pressure is given, the ZHD value can be computed.

From the second term in the bracket of Eq. 3.61, Saastamoinen (1972) determine the ZWD with an assumption that the partial pressure water vapor and temperature were decreased linearly with height. The final expression of ZWD is

$$\text{ZWD}_{\text{SAAS}} = 0.002277 \left(\frac{1225}{T_s} + 0.05\right) P_w \qquad (3.64)$$

where T_s is the surface temperature in °C and P_w is the partial water vapor in mbar.

3.3.2 The Hopfield Model

Hopfield (1969) developed a dual quartic zenith model of the refractivity with different quartics for the dry and wet atmospheric profiles using real data of surface measurements (pressure, temperature, and humidity) covering the whole Earth. This model assumes that the atmosphere is in hydrostatic equilibrium, which follows from the ideal gas law. The model also assumes the acceleration due to gravity and lapse rate in temperature is constant with height derived from a least-square fit to collected data. The model expresses the total delay in terms up to the fourth power of the refractive index. A representation of the dry and wet refractivity can be written as a function of height h above the surface by

$$N_j^{\text{Trop}}(h) = N_{j,0}^{\text{Trop}} \left(1 - \frac{h}{h_j}\right)^4 \qquad (3.65)$$

with total refractivity at surface of the Earth given as

$$N_{j,0}^{\text{Trop}} = N_{d,0}^{\text{Trop}} + N_{w,0}^{\text{Trop}} = \underbrace{k_1 \frac{P}{T_K}}_{\text{dry}} + \underbrace{k_2 \frac{P_w}{T_K} + k_3 \frac{P_w}{T_K^2}}_{\text{wet}} \qquad (3.66)$$

where j is the subscript for dry component (replace j by d) and wet component (replace j by w). N_j^{Trop} is the refractivity above the Earth surface, $N_{j,0}^{\text{Trop}}$ is the refractivity at the surface of the Earth, k_2 and k_3 are refraction constants

($k_2 = -12.96$ K mbar^{-1} and $k_3 = 5.718 \times 10^5$ K^2 mbar^{-1}), and h_j are the hydrostatic and wet thickness of atmospheric layer (m), respectively.

It assumes the atmosphere is a single polytrophic layer, thickness h_d was obtained by using global radiosonde data (Hofmann-Wellenhof et al. 2001):

$$h_d = 40{,}136 + 148.72\,T_K \qquad (3.67)$$

Unique values for h_d cannot be given because they depend on location and temperature. Figure 3.4 shows the thickness of polytrophic layers for the tropospheric. The effective troposphere heights are given as 40 km $\leq h_d \leq$ 45 km and 10 km $\leq h_w \leq$ 13 km for dry and wet components, respectively. The effective height for the wet component h_w is usually set to a default value of 11 km. Alternatively, Mendes and Langley (1998) found the relation between the surface temperature and the tropopause height denoted as H_T (in meters),

$$h_w = H_T = 7{,}508 + 0.002421\,\exp\left(\frac{T}{22.90}\right) \qquad (3.68)$$

Referring to Fig. 3.4, substitution of Eqs. 3.66 and 3.67 into the general Eq. 3.30 for the tropospheric path delay (ZPD) yields

$$\text{ZPD} = 10^{-6} N_{j,0}^{\text{Trop}} \int \left(1 - \frac{h}{h_j}\right)^4 ds \qquad (3.69)$$

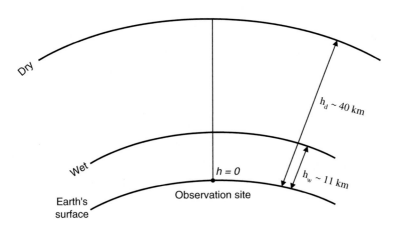

Fig. 3.4 Thickness of polytrophic layers for the troposphere adapted from Hofmann-Wellenhof et al. (2001)

The integral can be solved if the delay is calculated along the vertical direction and if the curvature of the signal path is neglected. Extracting the constant denominator, Eq. 3.69 becomes

$$\text{ZPD} = 10^{-6} N_{j,0}^{\text{Trop}} \frac{1}{h_j^4} \int_{h=0}^{h=h_j} (h_j - h)^4 \, dh \tag{3.70}$$

For an observation site on the Earth's surface (i.e., $h = h_s$) and after integration,

$$\text{ZPD} = 10^{-6} N_{j,0}^{\text{Trop}} \frac{1}{h_j^4} \left[-\frac{1}{5} (h_j - h)^5 \Big|_{h=h_s}^{h=h_j} \right] \tag{3.71}$$

The evaluation of the expression between the brackets gives the ZPD as follows:

$$\text{ZPD} = \frac{10^{-12}}{5} N_{j,0}^{\text{Trop}} \frac{1}{h_j^4} (h_j - h_s)^5 \tag{3.72}$$

where h_s is the height position of the receiver at site (in meters). If $h_s = 0$ as shown in Fig. 3.4, Eq. 3.72 can be rewritten as given by Hofmann-Wellenhof et al. (2001) and separating the hydrostatic and wet components, the total ZPD (in meters) is

$$\text{ZPD} = \frac{10^{-12}}{5} N_{j,0}^{\text{Trop}} h_j = \frac{10^{-6}}{5} \left[N_{d,0}^{\text{Trop}} h_d + N_{w,0}^{\text{Trop}} h_w \right] \tag{3.73}$$

The model in its present form does not account for an arbitrary elevation angle of the signal. Considering the line of sight, an obliquity factor must be applied for projecting the dependence of the zenith delays to the slant direction as a mapping function. Therefore, a slight variation of the Hopfield model contains an arbitrary elevation angle θ at the observation site using $1/\sin(\theta^2 + 6.25)^{1/2}$ as a mapping function for the hydrostatic component and $1/\sin(\theta^2 + 2.25)^{1/2}$ for the wet component. Hence, the total tropospheric delay at a zenith can be written as follows

$$\text{ZTD}_{\text{HOP}}(\theta) = \text{ZHD}_{\text{HOP}}(\theta) + \text{ZWD}_{\text{HOP}}(\theta) \tag{3.74}$$

where

$$\text{ZHD}_{\text{HOP}}(\theta) = \frac{10^{-6}}{5} \frac{77.64 \frac{P}{T_K}}{\sin \sqrt{(\theta^2 + 6.25)}} h_d \tag{3.75}$$

$$\text{ZWD}_{\text{HOP}}(\theta) = \frac{10^{-6}}{5} \frac{(-12.96 T_K) + 3.718 \times 10^5}{\sin \sqrt{(\theta^2 + 2.25)}} \left(\frac{e_s}{T_K^2} \right) h_w \tag{3.76}$$

3.3.3   The Modified Hopfield Model

The reference of station height on the Earth surface is inaccurate because of the
terrestrial points to be referred to a global frame. To overcome this limitation, the
atmospheric layer is considered to have azimuthally symmetry in ZTD estimation.
Therefore, the modified Hopfield model (Hofmann-Wellenhof et al. 2001) is refined
introducing the lengths of position vectors instead of height to correct the Hopfield
model for the determination of refractivity and denoting the earth's radius by R_E,
the corresponding lengths are $r_{hyd} = R_E + h_{hyd}$ and $r = R_E + h_{wet}$ as shown in
Fig. 3.5. R_E is taken as 6,378,137 meters in this paper. The empirical representation
of refractivity to the Modified Hopfield model, N_j as a function of height h above
the surface can be written as (Hofmann-Wellenhof et al. 2001),

$$N_j^{\text{Trop}}(r) = N_{j,0}^{\text{Trop}} \left(\frac{r_j - r}{r_j - R_E} \right)^4 \tag{3.77}$$

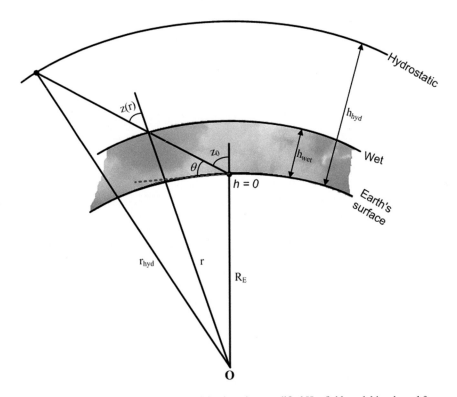

Fig. 3.5 Geometry for tropospheric path delay based on modified Hopfield model is adapted from
Hofmann-Wellenhof et al. (2001)

where for the hydrostatic refractivity component subscript j is replaced by hyd and for the wet refractivity component subscript j is replaced by wet, respectively. In the equation, N_j^{Trop} represent the refractivity above the earth surface and $N_{j,0}^{\text{Trop}}$ is the refractivity at the surface of the earth.

In this work, we corrected the refractivity by taking the dry $\left(Z_{\text{dry}}^{-1}\right)$ and wet $\left(Z_{\text{wet}}^{-1}\right)$ inverse compressibility factors into account for the determination of $N_{j,0}^{\text{Trop}}$ assuming that a nonideal gas represents the neutral atmosphere layer. Both the formula for Z_{dry}^{-1} and Z_{wet}^{-1} have been determined empirically by Owens (1967) as shown in Eqs. 3.6 and 3.7. By applying the ideal gas equation of state to the dry refractivity component in the Thayer equation (1974), $N_{\text{dry}} = k_1 \left(P_{\text{dry}}/T_{\text{K}}\right) Z_{\text{dry}}^{-1}$, the dry inverse compressibility factor $\left(Z_{\text{dry}}^{-1}\right)$ is eliminated and this term is changed to the hydrostatic term, $N_{\text{hyd}} = k_1 (P/T_{\text{K}})$. The refraction constant k_2 in the wet term of Eq. 3.66 is also corrected with a new constant k_2' as shown in Eq. 3.17. The total refractivity at the surface of the earth is then given as

$$N_{j,0}^{\text{Trop}} = N_{\text{hyd},0}^{\text{Trop}} + N_{\text{wet},0}^{\text{Trop}} = \underbrace{k_1 \frac{P}{T_{\text{K}}}}_{\text{hydrostatic}} + \underbrace{\left(\underbrace{k_2' \frac{P_{\text{wet}}}{T_{\text{K}}} Z_{\text{wet}}^{-1}}_{\text{dipole moment}} + \underbrace{k_3 \frac{P_{\text{wet}}}{T_{\text{K}}^2} Z_{\text{wet}}^{-1}}_{\text{dipole orientation}} \right)}_{\text{wet}} \quad (3.78)$$

where in the first term is hydrostatic refractivity in equilibrium state and the last term is the wet refractivity component. In Eq. 3.78 the 'dry' term has been replaced by 'hydrostatic' term.

Taking Eq. 3.78 for the hydrostatic delay and introducing mapping function $(1/\cos z)$, where zenith angle, $z(r) = 90° - \theta(r)$ is a variable and θ is the elevation angle at the observation site as shown in Fig. 3.5, the ZHD after applying the sine law can be expressed as

$$\text{ZHD} = \frac{10^{-6} N_{\text{hyd},0}^{\text{Trop}}}{\left(r_{\text{hyd}} - R_{\text{E}}\right)^4} \int_{r=R_{\text{E}}}^{r=r_{\text{d}}} \frac{r(r_{\text{hyd}} - r)^4}{\sqrt{r^2 - a^2}} \, dr \quad (3.79)$$

where the terms in the integral are constant except for r which is variable and $a = R_{\text{E}} \cos \theta$. Assuming the same model for the wet component, the corresponding formula is given by

$$\text{ZWD} = \frac{10^{-6} N_{\text{wet},0}^{\text{Trop}}}{\left(r_{\text{wet}} - R_{\text{E}}\right)^4} \int_{r=R_{\text{E}}}^{r=r_{\text{w}}} \frac{r(r_{\text{wet}} - r)^4}{\sqrt{r^2 - a^2}} \, dr \quad (3.80)$$

The integral in both equations can be solved by a series expansion of the integrand. Adopting the series expansion of Goad and Goodman (as cited in Hofmann-Wellenhof et al. 2001) the solution to the integral r_j is given as follows:

$$r_j = \left[R_E^2 \left(1 + \frac{h_j}{R_E} \right)^2 - a^2 \right]^{1/2} - \left[R_E^2 - a^2 \right]^{1/2} \tag{3.81}$$

Solutions of the total ZTD (in meters) as a function of θ, P, T, and H from Eqs. 3.79 and 3.80 can be expressed as (Suparta et al. 2008)

$$ZTD = 10^{-6} N_{j,0}^{\text{Trop}} \left[\begin{array}{l} 1 + 4a_j \frac{r_j^2}{2} + \left(6a_j^2 + 4b_j \right) \frac{r_j^3}{3} + 4a_j \left(a_j^2 + 3b_j \right) \frac{r_j^4}{4} \\ + \cdots \left(a_j^4 + 12a_j^2 b_j + 6b_j^2 \right) \frac{r_j^5}{5} + 4a_j b_j \left(a_j^2 + 3b_j \right) \frac{r_j^6}{6} \\ + \cdots b_j^2 \left(6a_j^2 + 4b_j \right) \frac{r_j^7}{7} + 4a_j b_j^3 \frac{r_j^8}{8} + b_j^4 \frac{r_j^9}{9} \end{array} \right] \tag{3.82}$$

$$a_j = -\frac{\sin \theta}{h_j} \quad \text{and} \quad b_j = -\frac{\cos^2 \theta}{2h_j R_E} \tag{3.83}$$

In general, Eq. 3.82 can be written as (Hofmann-Wellenhof et al. 2001)

$$ZTD(\theta, P, T, H) = 10^{-6} N_j^{\text{Trop}} \left(\sum_{k=1}^{9} \frac{\alpha_{k,j}}{k} r_j^k \right) \tag{3.84}$$

In Eqs. 3.82 and 3.83, the factor of 10^{-6} was corrected from 10^{-12} in Hofmann-Wellenhof et al. (2001: 115) to meet a consistency solution from Eqs. 3.79 and 3.80. In Eq. 3.83, h_j (in meters) represent h_{hyd} and h_{wet} are the effective height for the hydrostatic and wet components, respectively. In this work, h_h in Eq. 3.67 is used and the tropopause height or wet component (h_{wet}) is set to 11 km. The elevation angle is extracted from the GPS signals. In Eq. 3.84, k is the tropospheric layer.

Comparing the ZTD accuracies for both Hopfield and Saastamoinen models, the standard deviations of both models have very small difference of about 0.2 and 12.4 mm for the hydrostatic and wet components, respectively. Note that unlike the Hopfield and Saastamoinen models described earlier for the zenith delay, the Modified Hopfield model is also introduced for slant delays.

3.4 The Mapping Function

A mapping function is defined as the ratio of the electrical path length (also referred to as the delay) through the atmosphere at a geometric elevation, to the electrical path length in the zenith direction (Niell 2000). It is developed due to the tropospheric delay is the shortest in the zenith direction and becomes larger with increasing zenith angle. For GPS measurement of zenith PWV, the signal delay in each direction to each GPS satellite is not generally estimated individually. Instead, the individual delays are mapped from each individual satellite direction to a single zenith delay. This mapping method assumes that the delay is independent of azimuth. This assumption could never be made because of the significant increase in delay that is seen when the signal travels through much more of the atmosphere at lower elevations. Mapping functions account for the delay for individual satellite view and map them to the zenith direction.

Similar to the tropospheric models, many mapping functions have been proposed, such as Black (1978), Baby et al. (1988), Chao (1972), Davis et al. (1985), Herring (1992), Hopfield (1969), and Niell (1996). However, three mapping functions above are widely used because it included the hydrostatic and wet mapping functions. Those models of Davis, Herring, and Niell are called CfA-2.2 (Harvard–Smithsonian Center of Astrophysics), MTT (Massachusetts Institute of Technology, MIT Haystack Observatory), and new mapping functions, respectively. The CfA-2.2 mapping function (Davis et al. 1985) was designed to achieve sub-centimeter accuracy at 5 degrees elevations. The MTT mapping function (Herring 1992) can be used to represent the elevation angle dependence of the tropospheric delay with an RMS of less than 0.2 mm for elevation angles larger than 3 degrees. The last one is the new global mapping function by Niell (1996), namely the Niell Mapping Function (NMF). The NMF mapping functions almost similar to the MTT, which can be used for elevation angles down to 3 degrees.

Nowadays, the NMF mapping derived from Very Long Base Interferometry (VLBI) observations is the most widely used and known to be most accurate and easily implemented functions. Niell (1996) recognized that mapping functions like those of CfA-2.2, MTT, and Ifadis (1992) which all depend on surface temperatures. Unfortunately, the temperatures are much more variable in particular at higher altitudes both diurnally and on longer time scales, resulting in an error in the mapping. Therefore, NMF was developed to be independent of surface meteorological parameters. In this section, a simplest (cosecant) mapping function is introduced, then the Niell mapping function. The selection of this functional model is based on its ability to perform well in both low and high elevation and its independence meteorological parameters (Leick 1995).

3.4.1 The Cosecant Mapping Function

Foelsche and Kirchengast (2001) introduce a simple "geometric" mapping function (Fig. 3.6), where only the free parameter is an "effective height" of the atmosphere, corresponding to about the first two scales height above the surface. The simplest mapping function is the cosecant of the elevation angle that assumes both the curvature of the earth and the curvature of the path of the GPS signal propagating through the atmosphere can be approximated as plane surfaces. This is a reasonably accurate approximation only for high elevation angles with a small degree of bending.

To simplify Fig. 3.6, the value of ds/dr is defined to be the ratio of the slant straight-line ray path length within the effective height, S_{atm} ($S_{atm} = s$) to the H_{atm} itself.

$$ds/dr = S_{atm}/H_{atm} \qquad (3.85)$$

The above equation in other ways can be written in the form directly expressing the deviation from the simple cosecant law

$$\frac{ds}{dr} = \frac{1}{\cos z \, \frac{S_{flat}}{S_{atm}}} \qquad (3.86)$$

where dr is the difference in radius (distance to the center of the Earth) of the two layers, ds is the distance difference, z is the zenith angle at an arbitrary layer, S_{flat}

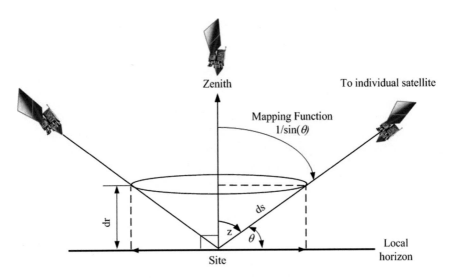

Fig. 3.6 Propagation of GPS signals approximated as a planar surface (Suparta 2008)

would be the ray path within H_{atm} in a flat (plane-parallel) atmosphere and ds/dr is the mapping function, later known as $m(z)$. Therefore, Eq. 3.86 is rewritten as

$$m(z) = \frac{1}{\cos z \frac{S_{flat}}{S_{atm}}} \tag{3.87}$$

For the planar atmosphere, assume that the Earth is flat and refractivity is constant, hence $S_{atm} \cong S_{flat}$. The cosecant mapping function becomes $1/\sin$ (*elevation*). As can be seen in Fig. 3.6, for an infinitesimal thin layer we have

$$m(z) = \frac{1}{\cos z} \equiv \sec z \rightarrow m(\theta) = \frac{1}{\sin \theta} \tag{3.88}$$

Because of the curvature of the atmosphere, this zenith angle change along the ray path. A simple mapping function in Eq. 3.88 is limited for use above ~ 60 degrees elevation. As the ratio of the thickness of the atmosphere to the radius of the earth decreases, the atmosphere appears more planar. This thickness varies with latitude and season. Thus, a possible proxy for the mapping function is some quantity that is a measure of the thickness of the atmosphere. The more complex mapping functions are based on the truncation of the continued fractions. This type of mapping functions includes Chao (1972), Davis et al. (1985), Marini (1972), and Niell (1996). The following is the description the Niell mapping function.

3.4.2 The Niell Mapping Function

Differing from most typical tropospheric delay models, Niell has developed hydrostatic and wet mapping functions with new forms and their combined use to reduce errors in geodetic estimation for observations as low as 3° in elevation. Although it has no parameterization in terms of actual meteorological conditions, they agree as well or better than mapping functions calculated from radiosonde profiles. In fact, when there is no information about the state of the atmosphere other than at the surface, the variation of the mapping function is found to be better modeled in terms of the seasonal dependence of the atmosphere, which is taken to be sinusoidal and in terms of the latitude and height above sea level of the site. The form adopted for this mapping function is the continued fraction of Marini (1972) with three constants but normalized to unity at the zenith as proposed by Herring (1992).

Marini (1972) was the first one to come up with the idea to use continued fractions. The most recent mapping functions are those of Herring (1992), Ifadis (1992), and Niell (1996), which used the continued fractions. Continued fractions have the advantage over models with Taylor's expansions like the Saastamoinen model because they fit for nearly the whole range of zenith angles (see Fig. 3.3).

The following is one type of the mapping function presented by Herring (1992) with continued fractions

$$M(z_0) = \frac{1 + a/(1 + b/(1 + c/(1 + \cdots)))}{\cos z_0 + \dfrac{a}{\cos z_0 + \dfrac{b}{\cos z_0 + \frac{c}{\cos z_0 + \cdots}}}} \tag{3.89}$$

where a, b, and c are mapping function coefficients to be determined. For low precision, the coefficients can be set to $a = b = c = 0$, which yields the cosecant model as introduced in Sect. 3.4.1.

Based on the continued fraction, Niell (1996) has developed hydrostatic and wet mapping functions with new forms and combinations. It is used to reduce errors in geodetic estimation to provide a better fit and give better accuracy over the latitude range 43° N to 75° N for observations down to 3 degrees elevation. The form adopted for Niell mapping function is the continued fraction of Marini (1972) with three a, b, and c constants in the following.

$$m_j(\theta) = \frac{1 + \dfrac{a_j}{1 + \dfrac{b_j}{1 + \frac{c_j}{1 + c_j}}}}{\sin \theta + \dfrac{a_j}{\sin \theta + \dfrac{b_j}{\sin \theta + c_j}}} \tag{3.90}$$

In addition to a latitude and seasonal dependence due to varying solar radiation, the hydrostatic mapping function should also be dependent on the height above the geoid of the point of observation because the ratio of the atmosphere "thickness" to the radius of curvature decreases with height. This does not apply to the wet mapping function since the water vapor is not in hydrostatic equilibrium and the height distribution of the water vapor is not expected to be predictable from the station height. The Niell mapping function for the hydrostatic (replace j by hyd) and wet (replace j by wet) components is of the following form

$$m_j(\theta) = \underbrace{m_h(\theta) + \Delta m(\theta)}_{\text{hydrostatic}} + m_{\text{wet}}(\theta) \tag{3.91}$$

For the hydrostatic mapping function, Niell (1996) adjusted the heights above the geoid. The sensitivity of the hydrostatic mapping function to the height above MSL was determined by beginning the ray-trace with nine different elevation angles between 3 and 90 degrees to give both the hydrostatic and wet path delays of each of the nine standard profiles with the values of pressure, temperature, and relative humidity at 1 and 2 km altitude. The height correction, $\Delta m(\theta)$, is given by

$$\Delta m(\theta) = \frac{dm(\theta)}{dh} h \tag{3.92}$$

where h is the height of the site above geoid in meters. The analytic height correction coefficients is taken to be

$$\frac{dm(\theta)}{dh} = \Delta m(\theta) = \frac{1}{\sin\theta} - f(\theta, a_{ht}, b_{ht}, c_{ht}) \qquad (3.93)$$

Here, $f(\theta, a_{ht}, b_{ht}, c_{ht})$ represents the three-term continued fraction expressed by Eq. 3.94 in the Marini mapping function,

$$f(\theta, a_{ht}, b_{ht}, c_{ht}) = \frac{1 + (a_{ht}/(1 + b_{ht}/(1 + c_{ht})))}{\sin\theta + (\sin\theta + a_{ht}/(\sin\theta + b_{ht}/(\sin\theta + b_{ht})))} \qquad (3.94)$$

In the above equation, the coefficients $a_{ht} = 2.53 \times 10^{-5}$, $b_{ht} = 5.49 \times 10^{-3}$, and $c_{ht} = 1.14 \times 10^{-3}$ was determined by least-square fits to the height corrections at the nine elevation angles. In these fittings, Niell used one for north latitudes of $15°$ for the whole year and two for north latitudes of $30°$, $45°$, $60°$, and $75°$, for the months January and July as tabulated in Cole et al. (1965).

Finally, the hydrostatic mapping function has normalized to yield a value of unity at the zenith and with a height correction, $\Delta m(\theta)$, which can be written as

$$m_{hyd}(\theta) = \frac{1 + (a/(1 + b/(1 + c)))}{\sin\theta + (a/\sin\theta + (b/\sin\theta + c))} + \left[\frac{1}{\sin\theta} - f(\theta, a_{ht}, b_{ht}, c_{ht})\right]h \quad (3.95)$$

For Niell, wet mapping function can be written as

$$m_{wet}(\theta) = \frac{1 + (a/(1 + b/(1 + c)))}{\sin\theta + (a/\sin\theta + (b/\sin\theta + c))} \qquad (3.96)$$

The coefficients a, b, and c in Eq. 3.89 were derived from temperature and relative humidity profiles of the U.S. Standard Atmosphere which is dependent on the latitude at North regions $15°$ (tropical), $30°$ (subtropical), $45°$ (midlatitude), 60 and $75°$ (subarctic) for the months of January (Winter) and July (Summer) and takes seasonal variations into account. Niell assumes that the Southern and Northern hemispheres are antisymmetric in time, i.e., the seasonal behavior is the same. In addition, he assumes the equatorial region is described by the $15°$ N latitude profile while the polar region is described by the $75°$ N latitude profile.

For the hydrostatic component, these coefficients are determined based on height, latitude, and DoY (day of year). However, for the wet mapping function, they depend only on the latitude. The coefficients for hydrostatic mapping function can be interpolated based on the parameter values extracted from Table 3.3 by the following interpolation rule.

Table 3.3 Coefficients of the hydrostatic mapping function

Coefficient	$\varphi = 15°$	$\varphi = 30°$	$\varphi = 45°$	$\varphi = 60°$	$\varphi = 75°$
a_{avg}	1.2769934×10^{-3}	1.2683230×10^{-3}	1.2465397×10^{-3}	1.2196049×10^{-3}	1.2045996×10^{-3}
b_{avg}	2.9153695×10^{-3}	2.9152299×10^{-3}	2.9288445×10^{-3}	2.9022565×10^{-3}	2.9024912×10^{-3}
c_{avg}	62.610505×10^{-3}	62.837393×10^{-3}	63.721774×10^{-3}	63.824265×10^{-3}	64.258455×10^{-3}
a_{amp}	0.0	1.2709626×10^{-5}	2.6523662×10^{-5}	3.4000452×10^{-5}	4.1202191×10^{-5}
b_{amp}	0.0	2.1414979×10^{-5}	3.0160779×10^{-5}	7.2562722×10^{-5}	11.723375×10^{-5}
c_{amp}	0.0	9.0128400×10^{-5}	4.3497037×10^{-5}	84.795348×10^{-5}	170.37206×10^{-5}

For latitude $|\varphi| \leq 15°$

$$F(\varphi, t) = F_{\text{avg}}(15°) + F_{\text{amp}}(15°) \cos\left(2\pi \frac{\text{DoY} - T_0}{365.25}\right) \tag{3.97}$$

For latitude range $15° \leq |\varphi| \leq 75°$,

$$F(\varphi, t) = F_{\text{avg}}(\varphi_i) + \left[F_{\text{avg}}(\varphi_{i+1}) - F_{\text{avg}}(\varphi_i)\right] \frac{\varphi - \varphi_i}{\varphi_{i+1} - \varphi_i}$$
$$+ \cdots \left\{F_{\text{amp}}(\varphi_i) + \left[F_{\text{amp}}(\varphi_{i+1}) - F_{\text{amp}}(\varphi_i)\right] \frac{\varphi - \varphi_i}{\varphi_{i+1} - \varphi_i}\right\} \tag{3.98}$$
$$\cos\left(2\pi \frac{\text{DoY} - T_0}{365.25}\right)$$

For latitude $|\varphi| \geq 75°$,

$$F(\lambda, t) = F_{\text{avg}}(75°) + F_{\text{amp}}(75°) \cos\left(2\pi \frac{\text{DoY} - T_0}{365.25}\right) \tag{3.99}$$

where φ is the user's latitude and the subscripts refer to the nearest tabular latitude, F is the mapping function calculated coefficients a, b, and c, separated into average values and amplitudes. T_0 is the day of a year for "maximum winter" which is set to 28 for Northern Hemisphere and 211 for the Southern Hemisphere. The average and amplitude values of the hydrostatic mapping function coefficients are listed in Table 3.3.

For the latitude $|\varphi| \leq 15°$,

$$F(\varphi, t) = F_{\text{avg}}(15°) + F_{\text{amp}}(15°) \cdot \cos\left(2\pi \frac{\text{DoY} - T_0}{365.25}\right) \tag{3.100}$$

For the latitude $|\varphi| \geq 75°$,

$$F(\lambda, t) = F_{\text{avg}}(75°) + F_{\text{amp}}(75°) \cdot \cos\left(2\pi \frac{\text{DoY} - T_0}{365.25}\right) \tag{3.101}$$

In case of the wet mapping function, the interpolation rule is also following the equation, but the average values for a_{wet}, b_{wet}, and c_{wet} are shown in Table 3.4.

For the latitude $|\varphi| \leq 15°$,

$$F(\varphi, t) = F_{\text{avg}}(15°) \tag{3.102}$$

Table 3.4 Coefficients of the wet mapping function

Coefficient	$\varphi = 15°$	$\varphi = 30°$	$\varphi = 45°$	$\varphi = 60°$	$\varphi = 75°$
a_{avg}	5.8021879×10^{-4}	5.6794847×10^{-4}	5.8118019×10^{-4}	5.9727542×10^{-4}	6.1641693×10^{-4}
b_{avg}	1.4275268×10^{-3}	1.5138625×10^{-3}	1.4572752×10^{-3}	1.5007428×10^{-3}	1.7599082×10^{-3}
c_{avg}	4.3472961×10^{-2}	4.6729510×10^{-2}	4.3908931×10^{-2}	4.4626982×10^{-2}	5.4736039×10^{-2}

For the latitude range $15° \leq |\varphi| \leq 75°$,

$$F(\varphi, t) = F_{\mathrm{avg}}(\varphi_i) + \left[F_{\mathrm{avg}}(\varphi_{i+1}) - F_{\mathrm{avg}}(\varphi_i) \right] \cdot \frac{\varphi - \varphi_i}{\varphi_{i+1} - \varphi_i} \qquad (3.103)$$

For the latitude $|\varphi| \geq 75°$,

$$F(\varphi, t) = F_{\mathrm{avg}}(75°) \qquad (3.104)$$

Conclusively, Tables 3.3 and 3.4 show the dependency of coefficients a, b, and c on temporal and spatial conditions for hydrostatic and wet mapping functions, respectively. To use the mapping function for any latitude, linear interpolation between the coefficients is required. Above 75° the same coefficients may be used as those at 75°. Between 15° N and 15° S, the coefficients may be considered constant. On this basis, the NMF mapping functions were estimated to be error by less than 4 mm from 12° down to 3° in comparison to the MTT mapping functions of Herring, but with smaller biases relative to ray traces than the MTT mapping functions.

References

Askne J, Nordius H (1987) Estimation of tropospheric delay for microwaves from surface weather data. Radio Sci 22:379–386

Baby HB, Golé P, Lavergnat J (1988) A model for the tropospheric excess path length of radio waves from surface meteorological measurements. Radio Sci 23:1023–1038

Bevis M, Businger S, Herring TA, Rocken C, Anthes RA, Ware RH (1992) GPS-meteorological: remote sensing of atmospheric water vapor using the global positioning system. J Geophys Res-Atmos 97:15787–15801

Bevis M, Businger S, Herring TA, Rocken C, Anthes RA, Rocken C, Ware RH, Chiswell S (1994) GPS meteorology: mapping zenith wet delays onto precipitable water. J Appl Meteorol 33(3):379–386

Black HD (1978) An easily implemented algorithm for the tropospheric range correction. J Geophys Res 38(B4):1825–1828

Bock O, Doerflinger E (2001) Atmospheric processing methods for high accuracy positioning with the global positioning system. Phys Chem Earth (A) 26(6–8):373–383

Brunner FK, Welsch WM (1993) Effect of the troposphere on GPS measurements. GPS World 4(1):42–51

Chao CC (1972) A model for tropospheric calibration from daily surface and radiosonde balloon measurement. jet propulsion laboratory, Pasadena, California. Tech Memorandum 391–350:16

Cole AE, Court A, Kantor AJ (1965) Model atmospheres. In: Valley S (ed) Handbook of geophysics and space environments. McGraw-Hill, New York (also available from NTIS as ADA056800)

Davis JL, Herring TA, Shapiro II, Rogers AEE, Elgered G (1985) Geodesy by radio interferometry: effects of atmospheric modeling errors on estimates of baseline length. Radio Sci 20:1593–1607

Foelsche U, Kirchengast G (2001) A new "geometric" mapping function for the hydrostatic delay at GPS frequencies. Phys Chem Earth PT A 26(3):153–157

Goad CC, Goodman L (1974) A modified Hopfield tropospheric refraction correction model. AGU annual fall meeting, San Francisco, CA (Abstract: EOS 55: 1106)

Gregorius TLH, Blewitt G (1999) Modeling weather fronts to improve GPS heights: a new tool for GPS meteorology? J Geophys Res-Sol Earth 104(7):15261–15279

Griffith DJ (ed) (1999) Introduction to electrodynamics. Prentice Hall, Singapore

Herring TA (1992) Modeling atmospheric delays in the analysis of space geodetic data. In: Proceedings of the symposium on refraction of transatmospheric signal in geodesy, The Netherlands 36, New series: 157–164

Hofmann-Wellenhof B, Lichtenegger H, Collins J (eds) (2001) Atmospheric effect on the global positioning system, theory and practice. Springer, Berlin

Hopfield HS (1969) Two-quartic tropospheric refractivity profile for correcting satellite data. J Geophys Res 74(18):4487–4499

Ifadis IM (1992) The excess propagation path of radio waves: study of the influence of the atmospheric parameters on its elevation dependence. Surv Rev 31:289–298

Langley RB, Kleusberg A, Teunissen PJG (1996) Propagation of the GPS signals. In: GPS for geodesy. Lecture notes in earth sciences. Springer, Berlin, pp 103–140

Lanyi G (1984) Tropospheric delay effects in radio interferometry. Telecommunications and data acquisition progress. Jet Propulsion Laboratory, Pasadena, pp 152–159

Leick A (ed) (1995) GPS satellite surveying. Wiley, New York

Marini JW (1972) Correction of satellite tracking data for an arbitrary tropospheric profile. Radio Sci 7(2):223–231

Mendes VB, Langley RB (1998) Tropospheric zenith delay prediction accuracy for airborne GPS high-precision positioning. In: Proceeding of the institute of navigation, 54th annual meeting 1998, pp 337–347

Niell AE (1996) Global mapping functions for the atmosphere delay at radio wavelengths. J Geophys Res 101(B2):3197–3246

Niell AE (2000) Improved atmospheric mapping functions for VLBI and GPS. Earth Planets Space 52:699–702

Owens JC (1967) Optical refractive index of air: dependence on pressure, temperature and composition. Appl Opt 6:51–58

Saastamoinen J (1972) Introduction to practical computation of astronomical refraction. Bul Geodes 106:383–397

Seeber G (1993) Satellite geodesy, foundations, methods and applications. Walter de Gruyter, Berlin

Smith WL (1966) Notes on the relationship between total precipitable water and surface dew point. J Appl Meteorol 5:726–727

Smith EK, Weintraub S (1953) The constants in the equation of atmospheric refractive index at radio frequencies. Proc Inst Radio Eng 41(8):1035–1037

Suparta W (2008) Characterization and analysis of atmospheric water vapour using ground-based GPS receivers at Antarctica. Ph.D. thesis, Faculty of Engineering, Universiti Kebangsaan Malaysia

Suparta W, Abdul Rashid ZA, Mohd Ali MA, Yatim B, Fraser GJ (2008) Observations of Antarctic precipitable water vapor and its response to the solar activity based on GPS sensing. J Atmos Sol-Terr Phys 70:1419–1447

Thayer GD (1974) An improved equation for the radio refractive index of air. Radio Sci 9 (10):803–807

Wallace JM, Hobbs PV (1997) Atmospheric science an introductory survey. Academic Press, New York

World Meteorological Organization (2000) General meteorological standards and recommended practices. Appendix A, WMO technical regulations 49, corrigendum

Yuan L, Anthes R, Ware R, Rocken C, Bonner W, Bevis M, Businger S (1993) Sensing climate change using the global positioning system. Geophys Res 98(D8):14925–14937

Chapter 4
Estimation of ZTD Using ANFIS

Abstract This chapter describes the estimation of zenith tropospheric delay (ZTD) using ANFIS technique followed by obtaining of ZTD data from GPS together with data location of dataset. This chapter also describes the method of analysis used in the development of ZTD estimation model. The results of ANFIS ZTD data will be compared to others models and will be discussed in this chapter.

Keywords Estimation model · Fuzzy clustering · Statistical analysis · *TroWav* · Antarctica · Equatorial region

4.1 Estimation of ZTD Model

One of the most important in the development of estimation model using ANFIS method is a selection of input variables. Input variables have the power to determine the design of the network architecture of ANFIS (Chang and Chang 2006). In addition, the size of the training is also very important, because it can affect the strength of the model in capturing the various characteristics of ZTD. Figure 4.1 shows the plot development model ZTD estimates with ANFIS method.

In this study, the testing process is used to ensure that the model has been trained to be able to capture all of the various characteristics of the targets and avoid excessive overfitting. Overfitting occurs when the model trained is too much, so the mapping of inputs–outputs will lose the ability to capture all of the characteristics that are not trained in the model. Therefore, the model built partially will lose the ability to characterize the parameter in the targeted area.

Furthermore, the input and output data used in this study were normalized by scaling between −1 and 1. The purpose of this process is to eliminate their dimensions and ensure that all variables receive the same treatment during the training of model (Nourami and Tomasi 2013). In addition, the normalized input

© The Author(s) 2016
W. Suparta and K.M. Alhasa, *Modeling of Tropospheric Delays Using ANFIS*,
SpringerBriefs in Meteorology, DOI 10.1007/978-3-319-28437-8_4

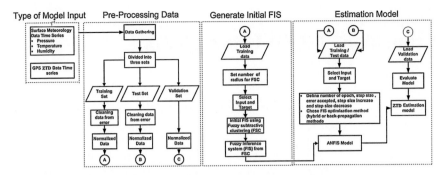

Fig. 4.1 The proposed method to determine the ZTD using ANFIS model

and output can accelerate the convergence process during the model training. A simple equation to normalize the data is expressed as follows:

$$Y_i = \left(2 \left(\frac{P_i - P_{\min}}{P_{\max} - P_{\min}} \right) - 0.5 \right) \tag{4.1}$$

where P_i and Y_i represent true and normalized values, respectively, and $P_{\max}$ and $P_{\min}$ are maximum and minimum values of the actual data, respectively.

4.1.1   Network Configuration Input for Estimation Model

Table 4.1 shows the model configurations that will be used in estimating the ZTD model. Configuration model consists of three input variables of a different parameter in surface meteorological data. This configuration is designed to avoid the limits relating to the availability of surface meteorological data for analyzing the suitability of each variable in improving the accuracy of the estimated ZTD model. Seven combinations of variables (A, B, C, D, E, F, and G) were designed to estimate the value of ZTD. The combinations that have been developed are from one input to three inputs, where the three variables are P, T, and H.

Table 4.1 Summary of configuration inputs used in the development of ANFIS model for estimation of ZTD

Combination	Input	Output
A	$P(t)$	$ZTD(t)$
B	$T(t)$	$ZTD(t)$
C	$H(t)$	$ZTD(t)$
D	$P(t), H(t)$	$ZTD(t)$
E	$P(t), T(t)$	$ZTD(t)$
F	$T(t), H(t)$	$ZTD(t)$
G	$P(t), T(t), H(t)$	$ZTD(t)$

4.1.2   Formation of a Fuzzy Inference System Using Fuzzy Clustering

In the development of ZTD model estimated using ANFIS method, the first step to be considered in addition to the variables and the size of the training is the development of a membership function (MF) and the rule base in FIS. In the development of the basic rules and membership functions normally using the grid partition, this method divides the data into exact rectangular subspaces by using axis-paralleled partition based on the dimension number of membership functions that have been set (Neshat et al. 2011). This method is often used to develop membership function and basic rules in the ANFIS method. However, it still has limitations in controlling the amount of very fast growing rules when increasing the number of input variables in the FIS. The number of rules that are too large will affect the future of computing time and the optimization of the rule parameters will not be easy. Therefore, the fuzzy clustering method is used in this study to facilitate the establishment and optimization of membership functions, and the basic rule in ANFIS model. With fuzzy logic clustering method, input–output data will be grouped into clusters. Information from these clusters would help in the FIS formation of Sugeno type, where one has the ability to model the behavior of input–output data with a less number of rules.

There are two methods of fuzzy logic clustering that will be used in this study, which are fuzzy clustering means (FCM) and fuzzy subtractive clustering (FSC). The two methods are adapted to meet the drawbacks in the ANFIS model. FCM is a data clustering algorithm in which the existence of each data point within a cluster is specified by a membership grade. This method was first introduced by Bezdek et al. (1984). Usually this method first will determine the total number of clusters of existing data. There are initial cluster centers to be determined and each data point will be given a membership grade. The iteration is then done to update cluster centers and the degree of membership of each data point to as close as possible to the centers of clusters. Iteration is done based on the reduction of criterion function (objective function) described as follows:

$$J = \sum_{k=1}^{n} \sum_{i=1}^{c} \mu_{ik}^{m} \| X_k - V_i \|^2, \quad 1 \leq m \leq \infty \tag{4.2}$$

where n is the number of data points, c is the number of clusters, X_k is the kth data point V_i is ith cluster center, μ_{ik} is the degree of membership function of the kth data in the ith cluster, and m is a constant greater than 1 (typically, $m = 2$) (Gomez-Skarmeta et al. 1999). Fuzzy partitioning is carried out through an iterative optimization of the objective function shown above, with the update of membership and cluster centers by

$$\mu_{ik} = \frac{1}{\sum_{j=1}^{c} \left(\frac{\|x_k - v_i\|}{\|x_k - v_j\|}\right)^{2/(m-1)}}, V_i = \sum_{k=1}^{N} (\mu_{ik})^m X_k / \sum_{k=1}^{N} (\mu_{ik}); 1 \leq i \leq c \quad (4.3)$$

Equation 4.3 began with a desired number of cluster c and an initial guess for each cluster center V_i, $i = 1,2,3, \ldots c$, and this iteration will stop when the procedure converges to a local minima. The output of FCM function is a list of the cluster center and several membership functions for each data. The information generated from FCM will greatly facilitate the development of initial FIS before training being done by ANFIS method, particularly in defining the membership functions to represent the fuzzy logic value of each clustering.

The FSC is one of the automated data-driven methods for constructing the primary fuzzy models (Chiu 1994) and an extension of the mountain clustering (Yager and Filev 1994). FSC assumes that each data point is a potential cluster center and calculates a measure based on the density of surrounding data points. The main process of subtractive clustering is that each data point is considered as a potential cluster center instead of a grid point. The density measure of data points at x_i is defined by the following equation:

$$D_i = \sum_{i=1}^{N} \exp\left(-\frac{\|x_i - x_j\|}{\left(\frac{r_a}{2}\right)^2}\right) \quad (4.4)$$

where r_a is a constant referred to as the neighborhood radius of a cluster center. After the density of each data point is selected, the point with the highest density potential D_{c1} and the first cluster center x_{c1} were chosen. Then, the density of each data point x_i can be updated by the following equation:

$$D_i = D_i - D_{c1} * \exp\left(-\frac{\|x_i - x_{c1}\|}{(r_b/2)^2}\right) \quad (4.5)$$

where $r_b = 1.5\,r_a$. After calculating the density for each data point is updated, the next cluster center x_{c2} is selected and all calculations of the density for data points to be updated again. This process is repeated until the number of generated cluster centers is sufficient. In this study, subtractive clustering algorithm uses the following samples parameter: the range of influence of the group, which marks the radius adjacent, r_a is 0.2–0.6, the acceptance ratio is 0.5 and the rejection ratio is 0.15, where the acceptance and rejection ratios show the conditions for accepting or rejecting data point to a cluster center, respectively. The parameter of each model was determined using the trial and error methods.

The difference between the FCM and the FSC methods lies in the way they determine the number of clusters. For FCM, initial value and clustering number should be fixed in order to achieve an optimal solution. On the other hand, the optimal solution in determining the clustering number through FSC is independent of the initial value of the cluster. The number of clusters generated will be used as the number of fuzzy rule's premise in the ANFIS model. Finally, the parameters of ANFIS model can be adapted much more efficiently.

4.1.3 Multi-layer Perceptron Network

Referring to Sect. 2.2 in Chap. 2, there are two types of ANN architecture: feed-forward and feedback neural networks. In this study, one of the existing ANN architectures will be applied to develop a comparison model to the ANFIS model. One of the structures that often used by researchers in ANN modeling is a multi-layer perceptron network (MLP). MLP is feedforward ANN that consists of three layers: input layer, hidden layer, and output layer. MLP is a modification of the standard linear single perceptron. A multi-layer perceptron is very useful to approach the classification function that maps the input vector to one or more classes, with optimizing the weights and thresholds for all nodes, the network can represent a variety of classification functions. In order to optimize the weight of the synapse, a supervised learning algorithm is typically used in the MLP networks. A typical feedforward constructed for MLP is shown in Fig. 4.2.

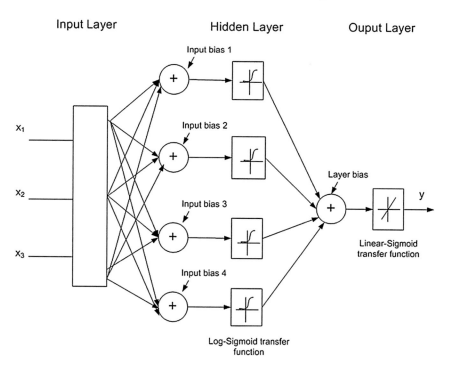

Fig. 4.2 The ANN model with three inputs, one hidden layer with four nodes, and one output (Suparta and Alhasa 2013)

In this study, MLP model is developed with a three-layer structure consisting of an input layer, one hidden layer, and output layer. To train the MLP network, the Levenberg–Marquadth learning algorithm (Yu and Wilamowski 2011) is applied to update the weight and bias values. Furthermore, each bipolar sigmoid activation and linear activation functions are used as an activation function or threshold functions in the hidden layer and output layer. In determining the MLP network topology, the number of hidden layer is selected from 3 to 15 nodes to determine the optimal number of nodes in the hidden layer. After several trials of nodes in the hidden layer are evaluated, the number of nodes that have been optimum results will be selected and applied to the model as a comparison method to ANFIS models.

4.1.4   Multiple Linear Regressions

The second method to be used for comparison of ANFIS model in this study is a multiple linear regression (MLR). As explained in the Sect. 2.3, the input variables of MLR will be P, T, and H. From Eq. 3.23, a least squares method is used to determine the values of intercept β_0 and regression coefficients, $\beta_1 \ldots \beta_m$. This method minimizes the sum of squares of the vertical distance from each data point on the line (Park et al. 2015). Furthermore, four statistical evaluation criteria are used to assess the different performance models, namely root mean square error (RMSE), coefficient of determination (R^2), mean absolute error (MAE), and percent error (PE).

The four of statistical analysis is described below.

1. RMSE

$$\text{RMSE} = \sqrt{\frac{\sum_{i=1}^{n}(P_i - O_i)}{n}} \tag{4.6}$$

2. Coefficient of determination (R^2)

$$R = \frac{E_O - E_s}{E_s} \tag{4.7}$$

where

$$E_O = \sum_{i=1}^{n}\left(P_i - \overline{P_i}\right)^2 \quad \text{and } E_s = \sum_{i=1}^{n}\left(P_i - O_i\right)^2 \tag{4.8}$$

The scale of the determination coefficient (R^2) is defined as follows:

$$\begin{array}{ll} \left| R^2 \right| > 0.64 & \text{Strong} \\ 0.36 < \left| R^2 \right| < 0.64 & \text{Moderate} \\ 0.16 < \left| R^2 \right| < 0.36 & \text{Low} \\ \left| R^2 \right| < 0.16 & \text{No correlation} \end{array}$$

3. MAE

$$\text{MAE} = \sum_{i=1}^{n} \frac{|P_i - O_i|}{n} \tag{4.9}$$

4. PE (Chaudhuri and Middey 2011)

$$\text{PE} = \frac{\langle |O_i - P_i| \rangle}{\langle O_i \rangle} \tag{4.10}$$

where P_i and O_i denote predictions of the ZTD (m) and the observation of GPS ZTD (m), respectively, and $\overline{P_i}$ represent the average of the ZTD prediction data (m) and the observation data of GPS ZTD (m). The symbol n denotes the number of data samples in the test cases.

4.2   Computation of ZTD from GPS Measurements

4.2.1   ZTD Data Processed from TroWav

If we recall Sect. 3.3 in Chap. 3, when the signal travels through inhomogeneous space, the code (P) and carrier (Φ) measurements are affected significantly. The code and the carrier phase observables can be expressed as follows (Dach et al. 2007; Klobuchar and Kunches 2003):

$$P_r^s = \rho_r^s + d\rho^s + c\left(dt^s - dT_r\right) + d_{ion}^s + d_{trop}^s + \varepsilon(P_{rx}) + \varepsilon(P_{mult}) \tag{4.11}$$

$$\Phi_r^s = \rho_r^s + d\rho^s + c\left(dt^s - dT_r\right) + \lambda_{cw}\, N_r^s - d_{ion}^s + d_{trop}^s + \varepsilon\left(\Phi_{rx}\right) + \varepsilon\left(\Phi_{mult}\right) \tag{4.12}$$

where
P_r^s is the pseudorange measurement by the GPS receiver to the satellite (m)
ρ_r^s is the true range or "geometrical" distance between satellite and receiver (m)
$d\rho^s$ is the orbit correction term (m)
c is the speed of light in free-space (3×10^8 ms^{-1})
dT_r is the receiver clock correction (m)
dt^s is the satellite clock correction (m)

d_{trop}^s is the tropospheric delay term (m)
d_{ion}^s is the ionospheric delay term (m)
$\varepsilon(P_{rx})$ is the error in pseudorange measurement due to receiver noise (m)
$\varepsilon(P_{mult})$ is the errors in pseudorange measurement due to multipath (m)
Φ_r^s is the carrier phase measurement by the GPS receiver to the satellite (m)
$\varepsilon(\Phi_{rx})$ is the error in phase measurement due to receiver noise (m)
$\varepsilon(\Phi_{mult})$ is the errors in phase measurement due to multipath (m)
λ_{cw} is the carrier wavelength (m cycles^{-1})
N_r^s is the integer carrier phase ambiguity between the GPS receiver and the
                   satellite (cycles)

The observation equations in Eqs. 4.11 and 4.12 are differed in two ways. The ionospheric refraction correction d_{ion} has the opposite sign. This means that the velocity of the carrier wave (the phase velocity) is actually increased ("advanced"), while velocity of the pseudorange (so-called "group velocity") is decreased. Therefore, the pseudorange is considered "delayed" and hence the range (or group) refractive index is greater than unity (Hofmann-Wellenhof et al. 2001). In addition, there is no ambiguity term $\lambda_{cw} N_r^s$ for pseudoranges. All errors except multipath and noise can be reduced using techniques such as single differencing, double differencing, and DGPS corrections (Lachapelle 2003).

From Eq. 4.11, the ionospheric delay (d_{ion}^s) and tropospheric delay (d_{trop}^s) play a crucial role in investigation of our atmosphere effects. On the other hand, the total atmospheric delay obtained from the GPS is caused by ionospheric delay and tropospheric delay, which can be expressed as

$$\text{Total Atmospheric Delay} = \text{Ionospheric Delay} + \text{Tropospheric Delay} \quad (4.13)$$

The ionospheric delay of GPS signals is given in Eq. 4.1, which is frequency-dependent (dispersive effect) and can be nearly eliminated by observations using a dual-frequency GPS receiver (Klobuchar and Kunches 2003). In other words, this allows the ionospheric effects to be largely removed by a linear combination of dual-frequency data. The magnitude of ionospheric delay on total atmospheric delay is considered larger and can be neglected. The remaining delay in the neutral atmosphere is the delay known as the total tropospheric delay. Based on this assumption, the tropospheric delay can be separated into the wet and the hydrostatic components. These separations were made in order to simplify the tropospheric delay modeling. Therefore, the total tropospheric delay can be expressed as

$$\text{Tropospheric Delay} = \text{Hydrostatic Delay} + \text{Wet Delay} \quad (4.14)$$

Thus, Eq. 4.1 in zenith direction can be written as ZTD = ZHD + ZWD.

With a reason that the GPS system is not registered with the International GNSS Service (IGS), and at the same time, we advance our knowledge in atmospheric studies using a GPS. Hence, we have proposed to develop a tropospheric water vapor program, so-called *TroWav* based on the empirical modeling concept as explained in Sect. 3.3 of Chap. 3. The program code was written in MATLAB. The algorithms of the *TroWav* include satellite elevation angle, ZTD, ZHD, ZWD, and mapping function calculations. The *TroWav* ZTD is generated with VMF1 mapping function (Suparta 2014). The measurement of these quantities in the zenith directions is further to obtain total water vapor column or vertical integrated water vapor. In the *TroWav* program, ZTD is calculated based on the modified Hopfield model, and the ZHD is calculated using the Saastamoinen model. In this work, the atmospheric layer is considered to have azimuthal symmetry in the ZTD calculation. Figure 4.3 shows the flowchart of ZTD processing in the *TroWav*. Accuracy of coordinates for the GPS stations is necessary in order to determine the ZHD exactly. In this case, the assessment of the accuracy coordinates used the International terrestrial reference frame, ITRF 2008. To cancel the residual tropospheric delay, a single differencing technique with baseline length below 10 km was implemented in the preprocessing for precise ZTD estimation.

As shown in Fig. 4.3, the hydrostatic Vienna Mapping Function (VMF1) (Boehm et al. 2006) was used to map the dependence of the zenith delays to the satellite elevation angle (Suparta et al. 2011). The cutoff elevation angle was set to 13°, which was able to minimize the multipath effects and to maintain the quality of the data. In the VMF1, Boehm et al. updated the "*b*" and "*c*" coefficients of the Marini where the continued fraction is formed as shown in the IMF equation

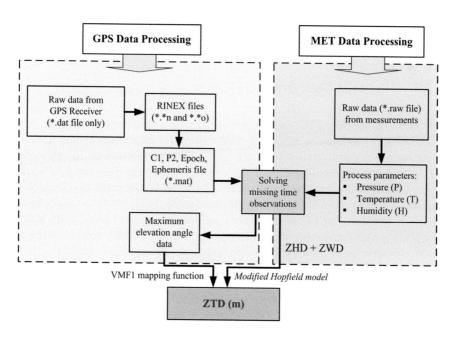

Fig. 4.3 Flowchart for computation of ZTD using GPS signals and surface meteorological measurements in the *TroWav* program

(Eq. 3.95) for the hydrostatic (index h) Niell mapping function, $m_h(\theta)$. However, the hydrostatic "a" coefficients are still valid for zero heights:

$$m_h(\theta) = m(\theta) + \Delta m(\theta)$$
$$= \frac{1 + (a/(1 + b/(1 + c)))}{\sin\theta + (a/\sin\theta + (b/\sin\theta + c))} + \left[\frac{1}{\sin\theta} - f(\theta, a_{ht}, b_{ht}, c_{ht})\right] h$$

$$(4.15)$$

where θ is the elevation angle (deg) and h is the height of the site above sea level, in meters.

For VMF1, the coefficient c as shown in Eq. 4.1 was now modeled to remove the systematic errors with taking consideration of European Centre for Medium-Range Weather Forecasts (ECMWF) instead of NWMs:

$$c = c_0 + \left[\left(\cos\left(\frac{\text{DoY} - 28}{365} 2\pi + \psi\right) + 1\right) \frac{c_{11}}{2} + c_{10}\right] (1 - \cos\varphi) \qquad (4.16)$$

where DoY is the day of the year and January 28 has been adopted as the reference epoch (Niell 1996), φ is the latitude (deg), and ψ specifies the northern (0) and southern hemispheres (π). For the equatorial region, all the coefficients above are set at half of the value for northern and southern. On the other hand, the ψ value was selected as $\pi/2$, where $c_0 = 0.062$, $c_{10} = 0.0015$, and $c_{11} = 0.006$. Detailed parameters for c_0, c_{10}, c_{11}, and ψ needed for computing the coefficient c in Eq. 4.1 of the hydrostatic mapping function can be found in (Boehm et al. 2006).

4.2.2   The ZTD Data from IGS

The effort to provide ZTD data is demand. In fact, this contribution can help scientists and geodesist in achieving their objectives, particularly in the field of Earth science research, multidisciplinary applications, and education. For example, IGS provides ZTD data of currently more than 300 globally distributed GPS stations. The ZTD data is stored into the database Crustal Dynamics Data Information System (CDDIS) NASA (ftp://cddis.gsfc.nasa.gov/pub/gps/products/troposphere/ZPD/). In this database, Gipsy–Oasis GPS software was used to generate the ZTD. The ZTD is calculated using a global mapping function (GMF) and processed at cutoff elevation angle of 7°, and resulted with a resolution of five minutes.

Beforehand, the analysis centers (ACs) such as the Center for Orbit Determination in Europe (CODE) of Astronomical Institute University of Bern (AIUB), Switzerland and GFZ Potsdam (Geo-ForshungsZentrum, Germany) provided the ZTD data from GPS. GFZ represents the weighted mean of the ZTD estimates contributed by 2–6 IGS analysis centers (Gendt 1998). They employed the cutoff elevation angle of 20° with the dry and wet Niell mapping functions for

CODE and GFZ analyses, respectively, in post-processing mode. Both ZTD products are available in 2-h intervals. However, the new ZTD data from AIUB can be found at the ftp://unibe.ch/aiub/CODE/, which is processed using a Bernese GNSS Software Version 5.3 at the cutoff elevation of 3° with a wet Vienna mapping function (VMF). The new resulting product is available with the hourly resolution.

Figure 4.4 shows the ZTD results for Scott Base (SBA: 77.9°S, 166.8°E and ellipsoid of 28.2 m) and McMurdo (MCM: 77.85°S, 166.67°E and a height of 98.01 m) in Antarctica compared to the IGS ZTD from CODE and GFZ, respectively, for the period of January 2004 at a 2-h resolution. In Fig. 4.4a, a small drift between *TroWav* ZTD and IGS ZTD (namely ZTDref) exhibited for two peaks, where *TroWav* ZTD reached its highest value the day before. This drift event is about 2 cm higher from their mean difference. The GPS and ZTD at both stations exhibit almost similar fluctuation compared to the IGS ZTD and shown a strong correlation ($r = 0.95$ at the 99 % confidence level). The mean difference values between *TroWav* ZTD and IGS ZTD range from 4.25 to 7.78 cm (see Fig. 4.4b). On the other hand, the ZTD average from *TroWav* at SBA and MCM for the period of January 2004 is 5.85 ± 0.68 cm higher compared to the IGS ZTD (ZTDref).

Looking at the ZTD data for the equatorial region, we compare the ZTD obtained in Singapore with the ZTD from IGS. The data collected from NASA CDDIS database of the year 2009 from the Nanyang Technology University of Singapore (NTUS: 1.35°N, 103.68°E and ellipsoid of 31.23 m) is processed.

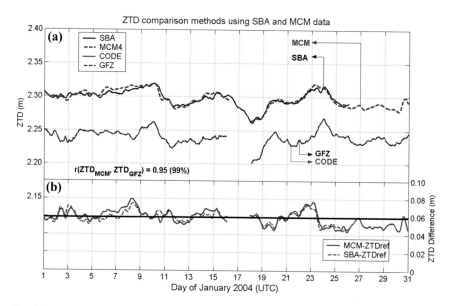

Fig. 4.4 The *TroWav* ZTD results from SBA and MCM compared to the IGS ZTD methods. The *solid horizontal line* in the bottom panel is the mean difference between ZTD at SBA and MCM respect to the IGS ZTD from CODE and GFZ, respectively

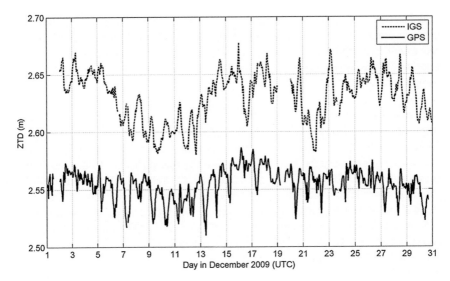

Fig. 4.5 Comparison results of estimated ZTD values through *TroWav* and IGS ZTD for NTUS station for the case of December 2009

Figure 4.5 shows the comparison of *TroWav* ZTD and IGS ZTD for the case of December 2009 at NTUS. A similar result as in Fig. 4.4 was found between ZTD generated from IGS with ZTD values generated from *TroWav*, where the IGS ZTD is greater than the *TroWav* method. More than this, the trend pattern of ZTD from *TroWav* follows the trend of IGS ZTD. The certain difference in value is possibly due to difference in the approach used. In the *TroWav* post-processing, the ZTD value with elevation angle below 30 degrees is removed to focus the result toward the zenith. This implies that all the errors that caused by the Earth's surface is eliminated. In this comparison, the difference between IGS ZTD with ZTD from *TroWav* is 2.95 % (or with an average of 7.55 cm).

4.3   Result of ZTD Estimated from ANFIS and GPS for Antarctica

The performance of estimation of ZTD from ANFIS models that was developed is tested under different conditions such as training dataset, set of test data, and validation dataset. For this purpose, we have selected five stations located in Antarctica with three stations along the Antarctica coast, which are SBA, Davis (DAV1: 68.58°S, 77.97°E and ellipsoid of 44.40 m), Syowa (SYOG: 69.00°S, 39.60°E and ellipsoid of 45 m), and two stations in the Antarctica Peninsula, which are Palmer (PALM: 64.80° S, 295.94°E and 31.23 m) and O'Higgins (OHI2: 63.32°S, 302.10°E and 33.10 m). At all stations, the surface meteorological data (*P, T,* and *H*) and IGS ZTD over the year 2010 are collected. We also tested the performance of ANFIS ZTD models developed for equatorial region, which are UKM Bangi (UKMB: 2.92°N, 101.77°E and ellipsoid of 35.64 m), Nanyang Technology University of Singapore (NTUS: 1.35°N, 103.68°E and ellipsoid of 31.23 m), and Universiti Malaysia Sabah Kota Kinabalu (UMSK: 6.03°N, 116.12°E and ellipsoid of 63.49 m).

Due to the unavailability of IGS ZTD data for SBA station, we did interpolation method to obtain the ZTD at the station (Suparta and Alhasa 2015). Two stations nearby the SBA based on the geographical coordinates is selected and used as a reference in the interpolation method, which are McMurdo (MCM4) and Dumont d'Urville (DUM1: 66.66°S, 140.00°E and height −1.33 m). From the smallest average error (bias) analysis between SBA-MCM4 and SBA-DUM1, coordinates based on longitude have the smallest average of −0.432 mm compared to latitude (−1.020 mm) and height (22.357 mm) in estimating the ZTD value. Figure 4.6 shows the comparison results of ZTD interpolation data for SBA, MCM4, and DUM1. The ZTD variation of ZTD at SBA was obtained very close and it has a similar variation to both stations with PE of 0.02 and 1.46 % for MCM4 and DUM1, respectively. The ZTD for SBA is 0.032 m lower at DUM1 and 0.012 m higher at MCM4. To differentiate the pattern, the ZTD variation between the three stations in Fig. 4.6 is shifted to a value of 0.15 and −0.15 m for DUM1 and MCM4, respectively.

For testing the model performance, one-year data at five selected stations is collected with aiming to cover all the various features of ZTD in the existing area. A large dataset will include all cases that occurred in the targeted area and it is important to build a model with artificial neural network method or ANFIS (Chang and Chang 2006). Figure 4.7 shows the results of each comparison between

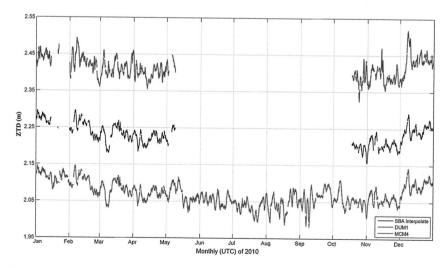

Fig. 4.6 Linear interpolation of ZTD results for SBA station. The ZTD values for DUM1 and MCM4, respectively, are shifted to 0.15 m and −0.15 m to distinguish the ZTD patterns

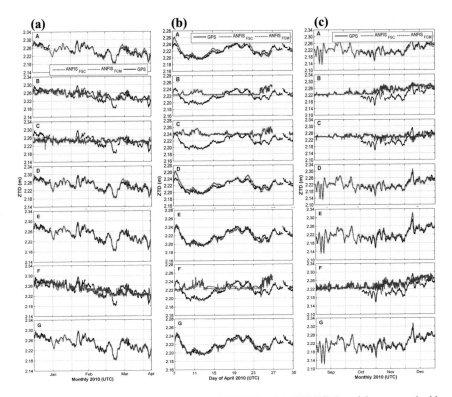

Fig. 4.7 Variation of ZTD estimated from ANFIS FSC and ANFIS FCM models compared with GPS ZTD data at SBA station during the **a** training, **b** testing, and **c** verification for (*A*) *P*, (*B*) *T*, (*C*) *H*, (*D*) *P* and *H*, (*E*) *P* and *T*, (*F*) *T* and *H*, and (*G*) *P*, *T*, and *H*

estimated ZTD using the ANFIS FSC and ANFIS FCM models with seven possibilities of input and compared with GPS ZTD at SBA station during the period of training, testing, and validation. The dashed lines (blue and red) in Fig. 4.7 represent the estimated value of ZTD using both ANFIS models and the black, solid line represents the GPS ZTD at SBA station. The period of training was from January until the first week of April 2010, while the period of testing and validation process was from 7 to 30 April 2010 and September to December 2010, respectively. The blank shown in the figure indicated no GPS or meteorological data recorded at that time. For the whole process, the ANFIS model that was trained using hybrid learning algorithm provides satisfactory results in which both ANFIS FSC and ANFIS FCM models are capable of fully capturing the characteristics of the ZTD pattern at SBA station. Furthermore, the results also show that the model estimated using a combination of three input variables (*P*, *T*, and *H*) has the best performance compared with other combinations of input variables than the only using *P*, *T*, or *H*, or *P* and *T*, or *P* and *H*, or, *T* and *H*. This can be seen from the figure that ZTD variation in the panel *B*, *C*, and *F* shows tough to follow the pattern of GPS ZTD. However, a combination of two input variables P and T can also be an alternative model when the surface meteorological data is limited. It has a good performance, where the coefficient of determination (R^2) is greater than 0.90.

Figure 4.8 shows the relationship between ZTD from ANFIS FSC and ANFIS FCM models and GPS ZTD from Fig. 4.7. Although the figure revealed any systematic bias distribution in estimating the ZTD and the random error for each input combination during the validation process, there found strong relationships with a positive linear trend. In addition, the figure shows that the combination of all inputs is a good model. However, the estimated ZTD without P, which use either T or H alone or a combination of both, will produce a output that does not capture the pattern of GPS ZTD. On the other hand, adding an input variable, particularly

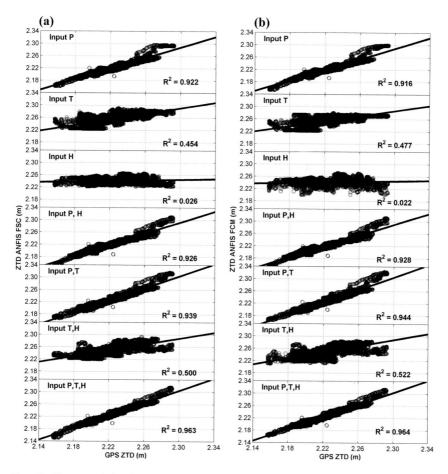

Fig. 4.8 Cross-correlation between ANFIS ZTD model and GPS ZTD data of Fig. 4.7 during the validation period for **a** ANFIS FSC and **b** ANFIS FCM

P into the ANFIS model, will improve the accuracy of the estimated ZTD using
ANFIS in Antarctica (Suparta and Alhasa 2015).

Similar results were also obtained for two other stations, DAV1 and SYOG
stations. For these two stations, the periods of training, testing, and validation are
the same, which are in January to mid-July 2010, mid-July to August 2010, and
September to December 2010, respectively. Figures 4.9 and 4.10 show the results
of ZTD values and any systematic bias in their distribution at DAV1 station, and for
SYOG station, it is shown in Figs. 4.11 and 4.12, respectively. Referring to the four
panels, a similar result to that of SBA was obtained for DAV1 and SYOG stations,
where the R^2 was greater than 0.90. The input with *T* only (*B*), *H* only (*C*), and
combination of *T* and *H*(*F*) shows a similar problem as in SBA that the model
cannot capture the characteristic of GPS ZTD. These inputs with no strong correlation from Figs. 4.9 and 4.11 are excluded from Figs. 4.10 and 4.12. For the input

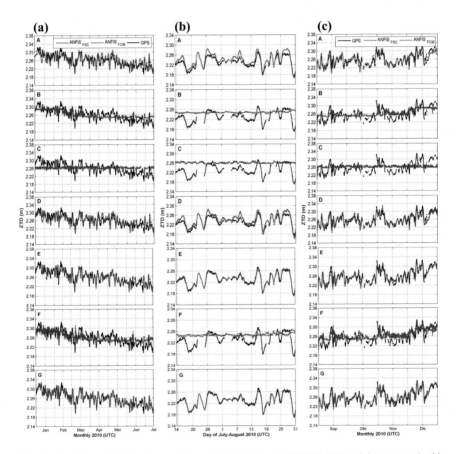

Fig. 4.9 Variation of ZTD estimated from ANFIS FSC and ANFIS FCM models compared with
GPS ZTD data at DAV1 station during the **a** training, **b** testing, and **c** verification for (*A*) *P*, (*B*) *T*,
(*C*) *H*, (*D*) *P* and *H*, (*E*) *P* and *T*, (*F*) *T* and *H*, and (*G*) *P*, *T*, and *H*

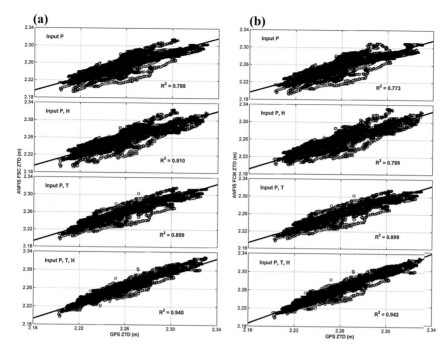

Fig. 4.10 Cross-correlation between ZTD ANFIS model and GPS ZTD data of Fig. 4.9 during the validation period for **a** ANFIS FSC and **b** ANFIS FCM

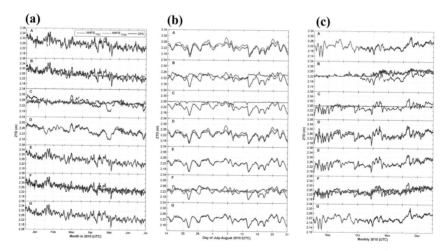

Fig. 4.11 Variation of ZTD estimated from ANFIS FSC and ANFIS FCM models compared with GPS ZTD data at SYOG station during the **a** training, **b** testing, and **c** verification for (*A*) *P*, (*B*) *T*, (*C*) *H*, (*D*) *P* and *H*, (*E*) *P* and *T*, (*F*) *T* and *H*, and (*G*) *P*, *T*, and *H*

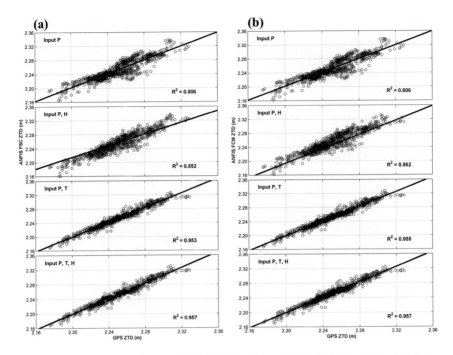

Fig. 4.12 Cross-correlation between ZTD ANFIS model and GPS ZTD data of Fig. 4.11 during the validation period for **a** ANFIS FSC and **b** ANFIS FCM

with a strong correlation, the distributions of bias and random errors are found more variable, especially for the estimation model that uses a combination of input variables P and H. For the overall results obtained with ANFIS models, the estimations of ZTD at both stations with combination input variables of P and T were satisfactory.

Furthermore, the result of two other stations in the Antarctic Peninsula, PALM, and OHI2 is depicted in Figs. 4.13, 4.14, 4.15, and 4.16, respectively. Referring to Fig. 4.13, the pattern of ANFIS ZTD model was agreed very well with the pattern of observed GPS ZTD at PALM station. In contrast, Fig. 4.15 shows the unsatisfactory results of ANFIS ZTD model, which does not fully capture the characteristics of the GPS ZTD at OHI2 station. The distribution bias and random error shown in Figs. 4.14 and 4.16 vary compared with the results obtained in the stations located in the Antarctica coast. As shown in both figures, this condition is found in the annual weather of Antarctica Peninsula, where the atmospheric environments are strongly influenced by the relationship between cyclone, temperature, and sea ice deep and continuous transition difference between maritime climate patterns and continental climate regime (Suparta et al. 2009). The atmospheric environment condition of OHI2 station is more humid with higher pressure depression compared with other stations. Therefore, the atmospheric environment condition at each

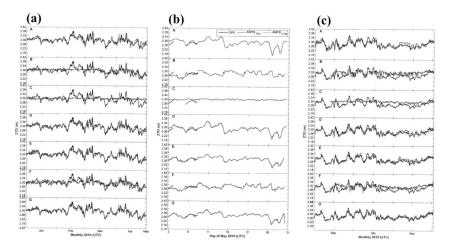

Fig. 4.13 Variation of ZTD estimated from ANFIS FSC and ANFIS FCM models compared with GPS ZTD data at PALM station during the **a** training, **b** testing, and **c** verification for (*A*) *P*, (*B*) *T*, (*C*) *H*, (*D*) *P* and *H*, (*E*) *P* and *T*, (*F*) *T* and *H*, and (*G*) *P*, *T*, and *H*

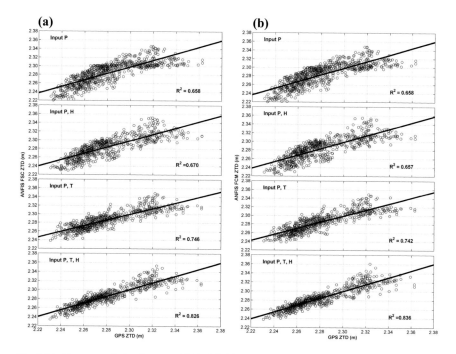

Fig. 4.14 Cross-correlation between ZTD ANFIS model and GPS ZTD data of Fig. 4.13 during the validation period for **a** ANFIS FSC and **b** ANFIS FCM

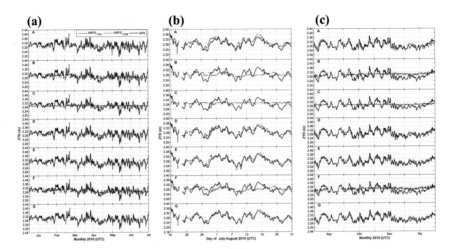

Fig. 4.15 Variation of ZTD estimated from ANFIS FSC and ANFIS FCM models compared with GPS ZTD data at OHI2 station during the **a** training, **b** testing, and **c** verification for (*A*) *P*, (*B*) *T*, (*C*) *H*, (*D*) *P* and *H*, (*E*) *P* and *T*, (*F*) *T* and *H*, and (*G*) *P*, *T*, and *H*

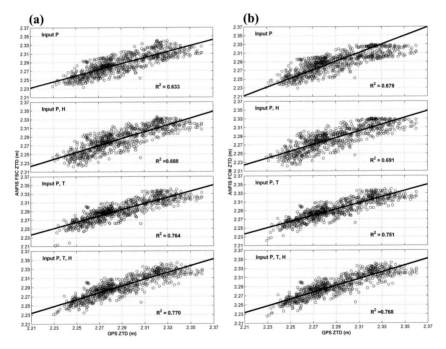

Fig. 4.16 Cross-correlation between ZTD ANFIS model and GPS ZTD data of Fig. 4.15 during the validation period for **a** ANFIS FSC and **b** ANFIS FCM

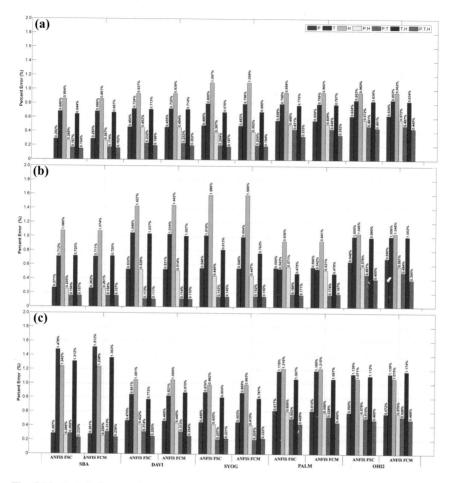

Fig. 4.17 Statistical comparison of percentage error (*PE*) between GPS ZTD and ANFIS ZTD model at five selected stations during the **a** training, **b** testing, and **c** validation

station will affect the ability of the model to capture all the characteristics of ZTD in the targeted area.

The results of other statistical analysis such as RMSE, MAE, and PE to evaluate the model performance at five selected stations in Antarctica during the training, testing, and validation sets are shown in Fig. 4.17 and Table 4.2. From the results, the estimation of ZTD from ANFIS FCM and ANFIS FSC models by adding a new input variable into the configuration model would improve the accuracy of ZTD values. These results are clearly visible where the error occurred in the model will gradually decrease during the addition of new inputs such as relative humidity and temperature.

Based on the performance, results obtained from each station showed that the ability of both ANFIS models will increase by adding the appropriate input

Table 4.2 Statistical comparison between ZTD obtained from GPS and ANFIS models with seven input networks at five selected stations in Antarctica during the process of training, testing, and validation. RMSE and MAE units are in meter, respectively

Site	Input	Training				Testing				Validation			
		ANFIS FSC		ANFIS FCM		ANFIS FSC		ANFIS FCM		ANFIS FSC		ANFIS FCM	
		RMSE	MAE	RMSE	MAE	RMSE	MAE	RMSE	MAE	RMSE	MAE	RMSE	MAE
SBA	P	0.008	0.007	0.008	0.007	0.007	0.006	0.007	0.006	0.008	0.006	0.008	0.006
	T	0.019	0.015	0.019	0.015	0.019	0.016	0.019	0.016	0.038	0.033	0.038	0.033
	H	0.024	0.019	0.024	0.019	0.028	0.024	0.028	0.024	0.033	0.027	0.033	0.027
	P, H	0.008	0.006	0.008	0.006	0.007	0.006	0.007	0.006	0.007	0.006	0.007	0.006
	P, T	0.005	0.004	0.005	0.004	0.004	0.003	0.004	0.004	0.008	0.007	0.009	0.007
	T, H	0.018	0.014	0.018	0.015	0.019	0.016	0.019	0.016	0.033	0.029	0.034	0.030
	P, T, H	0.004	0.003	0.005	0.004	0.004	0.003	0.004	0.003	0.006	0.005	0.007	0.005
DAVA1	P	0.013	0.010	0.013	0.010	0.013	0.012	0.013	0.012	0.013	0.011	0.013	0.011
	T	0.020	0.016	0.020	0.016	0.029	0.023	0.029	0.023	0.023	0.019	0.023	0.019
	H	0.026	0.021	0.026	0.021	0.038	0.032	0.038	0.032	0.029	0.024	0.029	0.024
	P, H	0.013	0.010	0.013	0.010	0.013	0.012	0.013	0.011	0.012	0.010	0.012	0.010
	P, T	0.007	0.005	0.007	0.005	0.003	0.003	0.003	0.003	0.009	0.007	0.009	0.007
	T, H	0.020	0.016	0.020	0.016	0.029	0.023	0.029	0.022	0.022	0.017	0.024	0.020
	P, T, H	0.006	0.005	0.006	0.005	0.003	0.002	0.003	0.002	0.007	0.006	0.007	0.006
SYOG	P	0.014	0.011	0.014	0.011	0.014	0.012	0.014	0.012	0.013	0.010	0.013	0.010
	T	0.024	0.018	0.023	0.018	0.030	0.022	0.029	0.022	0.025	0.020	0.028	0.022
	H	0.031	0.025	0.031	0.025	0.041	0.035	0.041	0.035	0.028	0.022	0.028	0.022
	P, H	0.011	0.009	0.012	0.009	0.012	0.010	0.012	0.010	0.012	0.009	0.011	0.009
	P, T	0.006	0.005	0.006	0.005	0.004	0.003	0.004	0.003	0.007	0.005	0.007	0.005
	T, H	0.020	0.015	0.020	0.015	0.024	0.018	0.023	0.017	0.023	0.018	0.022	0.018
	P, T, H	0.006	0.004	0.006	0.004	0.004	0.003	0.004	0.003	0.006	0.005	0.006	0.005

(continued)

Table 4.2 (continued)

Site	Input	Training				Testing				Validation			
		ANFIS FSC		ANFIS FCM		ANFIS FSC		ANFIS FCM		ANFIS FSC		ANFIS FCM	
		RMSE	MAE	RMSE	MAE	RMSE	MAE	RMSE	MAE	RMSE	MAE	RMSE	MAE
PALM	P	0.017	0.013	0.017	0.013	0.016	0.013	0.016	0.013	0.017	0.014	0.017	0.014
	T	0.024	0.018	0.024	0.018	0.016	0.012	0.016	0.012	0.032	0.027	0.032	0.027
	H	0.029	0.022	0.029	0.022	0.024	0.021	0.024	0.021	0.032	0.028	0.032	0.028
	P, H	0.016	0.011	0.016	0.012	0.015	0.013	0.014	0.012	0.016	0.014	0.017	0.014
	P, T	0.013	0.010	0.013	0.010	0.005	0.004	0.006	0.004	0.015	0.012	0.014	0.011
	T, H	0.023	0.018	0.023	0.018	0.014	0.011	0.014	0.011	0.029	0.024	0.029	0.024
	P, T, H	0.011	0.008	0.011	0.008	0.006	0.004	0.006	0.004	0.012	0.010	0.012	0.010
OHI2	P	0.018	0.014	0.018	0.015	0.018	0.015	0.019	0.016	0.017	0.014	0.017	0.013
	T	0.025	0.020	0.025	0.020	0.028	0.023	0.028	0.023	0.030	0.026	0.030	0.026
	H	0.028	0.022	0.028	0.022	0.029	0.024	0.029	0.024	0.029	0.025	0.029	0.025
	P, H	0.016	0.013	0.017	0.013	0.016	0.013	0.016	0.013	0.017	0.013	0.017	0.013
	P, T	0.014	0.011	0.014	0.011	0.013	0.011	0.014	0.011	0.015	0.012	0.015	0.012
	T, H	0.025	0.019	0.025	0.019	0.027	0.023	0.028	0.023	0.030	0.025	0.031	0.027
	P, T, H	0.013	0.010	0.013	0.010	0.012	0.009	0.011	0.009	0.014	0.011	0.014	0.011

variables. Furthermore, the ability of the model in estimating the ZTD value in Antarctica will increase when adding the temperature parameter rather than the relative humidity. This can be seen in Table 4.2, where the combination input variables of P and T have the second lowest error behind the three input variables (P, T, and H). As demonstrated in Fig. 4.17 by PE, the results also show that the ability of ANFIS FCM model is comparable with ANFIS FSC model, whether in the training set, testing, or validation processes.

From scatterplot between the GPS ZTD and ZTD from ANFIS, roughly, one input variable of using surface pressure for Antarctica is capable to develop a ZTD model. The percentage error of this model is below 1 % and a strong correlation is obtained for the use of the data with a maximum of a 1-h interval, and the modest correlation when using the data with a 3-h resolution. However, the best implementation of the ANFIS model to estimate ZTD in Antarctica is done by using two input variables, P and T.

4.4 Analysis Results of ANFIS ZTD in the Equatorial Region

As mentioned in the previous section, three stations namely UKMB, NTUS, and UMSK in the equatorial region are selected to test the performance of ANFIS ZTD model. A similar statistical analysis as in Antarctica was used to evaluate the model performance during different conditions of training, testing, and validation.

Figure 4.18 shows the results of ANFIS ZTD and their comparison with GPS ZTD at UKMB station. For UKMB, the training set period was from January until May and six of the first day of September in 2009. For the test set model, it is trained within the period of September 7–November 30, 2009, while the set validation period was given for March, August, October, and December 2009. According to the results obtained from Fig. 4.18, it can be seen that the pattern of the ZTD estimated from ANFIS FSC and ANFIS FCM models in this area differs to those found in the Antarctica. However, UKMB station has successfully captured the patterns of observed GPS ZTD that uses a combination of three input variables P, T, and H. The other combined model either using P, or P and H, or P and T was unsuccessful to follow the pattern of GPS ZTD. However, the combination of input variables T and H is seen as an alternative model to estimate the ZTD.

The difference compared to the Antarctica region is probably because of differences in the latitude and altitude of a concerned area, where the altitude affect changes in the earth's atmospheric pressure. The increase in altitude will cause low pressure in the atmosphere and vice versa (Jin et al. 2007). In fact, Antarctica region has a lower pressure value than the equatorial region. In addition, the surface pressure greatly affects the results of both ANFIS models in estimating the ZTD in the Antarctica (Suparta and Alhasa 2015). For the equatorial region, particularly Malaysia, the relative humidity is shown which significantly affects the results than

(a) **(b)** **(c)**

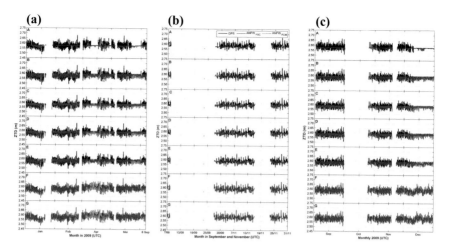

Fig. 4.18 Variation of ZTD estimated from ANFIS FSC and ANFIS FCM models compared with GPS ZTD data at UKMB station during the **a** training, **b** testing, and **c** verification for (A) P, (B) T, (C) H, (D) P and H, (E) P and T, (F) T and H, and (G) P, T, and H

the surface pressure. It can be seen that the correlation coefficient (r) between the input variables and GPS ZTD is 0.428 and r values for surface pressure and temperature are -0.120 and -0.147, respectively.

In order to see any systematic bias distribution and relative size of the random error during the validation period at all combination inputs, Fig. 4.19 shows the relationship between ANFIS ZTD models and GPS ZTD with all combination inputs. From the figure, it is shown that the combination of variables (P) and two combinations of input variables (P and T, and P and H) in estimating the relative size of ZTD found large random errors and systematic errors. The R^2 of the three variables of combined model is found to be lower than 0.40 or weak relationship. Similar to the previous finding, the three combinations of input variables P, P and T, and P and H did not succeed in capturing all the characteristics of GPS ZTD in the concerned area. On the other hand, the model with three input variables (P, T, and H) found a strong relationship, where the R^2 value is more than 0.999.

A similar result was obtained for the two other stations located in Singapore (NTUS station) and Borneo (see Figs. 4.20 and 4.21). Referring to Fig. 4.20, the different resolutions of data indicated different backgrounds of the graph. The solid and the thin graphs showed the data with 1-h and 3-h intervals, respectively. The blank data in time series was also due to no GPS or meteorological data. From Fig. 4.21, it shows that the period of training period was January, February, April, and mid of May (1–16 May of 2009), the testing period was from May 17 to 31 until June 30, 2009, while the validation period was March, July, August, and December 2009. For the overall results, input model with three variables (P, T, and H) was the

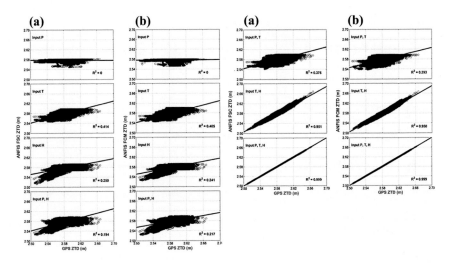

Fig. 4.19 Cross-correlation between ZTD ANFIS model and GPS ZTD data of Fig. 4.18 during the validation period for **a** ANFIS FSC and **b** ANFIS FCM

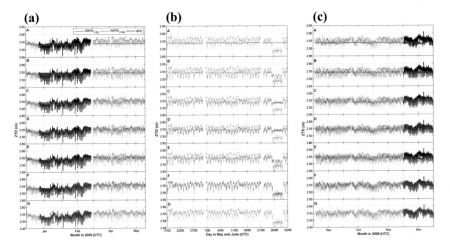

Fig. 4.20 Variation of ZTD estimated from ANFIS FSC and ANFIS FCM models compared with GPS ZTD data at NTUS station during the **a** training, **b** testing, and **c** verification for (*A*) *P*, (*B*) *T*, (*C*) *H*, (*D*) *P* and *H*, (*E*) *P* and *T*, (*F*) *T* and *H*, and (*G*) *P*, *T*, and *H*

best performance in estimating the ZTD value at NTUS. The accuracies of combinational inputs for *T* and *H* are also behind the input values *P*, *T*, and *H*.

The results for UMSK station are shown in Figs. 4.21 and 4.23. The period of the training period was in May, July, and September 2, 2012, testing period was 28 days from 3 to 31 September 2012, while the validation period was July and August as shown in Fig. 4.22. Based on Figs. 4.20, 4.21, 4.22, and 4.23, similar

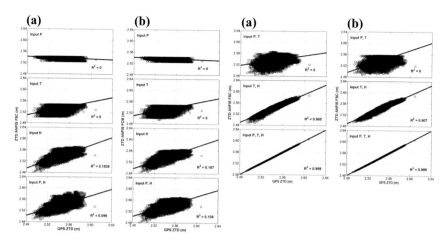

Fig. 4.21 Cross-correlation between ZTD ANFIS model and GPS ZTD data of Fig. 4.20 during the validation period for **a** ANFIS FSC and **b** ANFIS FCM

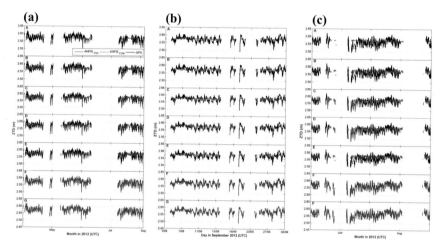

Fig. 4.22 Variation of ZTD estimated from ANFIS FSC and ANFIS FCM models compared with GPS ZTD data at UMSK station during the **a** training, **b** testing, and **c** verification for (*A*) *P*, (*B*) *T* (*C*) *H* (*D*) *P* and *H*, (*E*) *P* and *T*, (*F*) *T* and *H*, and (*G*) *P*, *T*, and *H*

results are obtained for NTUS and UMSK stations. The ANFIS FSC and ANFIS FCM models were successful to estimate ZTD values with three input variables (*P*, *T*, and *H*). By these three input variables, the ZTD model shows a relatively small error of size distribution and the R^2 value was greater than 0.99 (see Figs. 4.21 and 4.23).

The other detailed statistics (RMSE, MAE, and PE) in performing the estimation of the ZTD values between ANFIS ZTD models and observed GPS ZTD data for

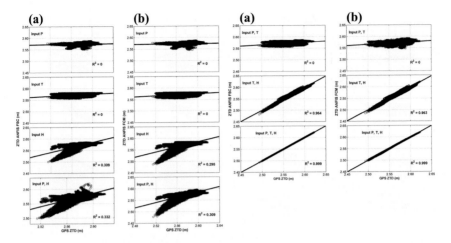

Fig. 4.23 Cross-correlation between ANFIS ZTD model and GPS ZTD from Fig. 4.22 during the validation period for **a** ANFIS FSC and **b** ANFIS FCM

Malaysia station (UKMB and UMSK) and Singapore station (NTUS) are compiled in Table 4.3 and Fig. 4.24. As shown in Table 4.3, the lowest error at all three stations using ANFIS FSC and ANFIS FCM models that use a combination of three input variables (P, T, and H) either during the training, testing, and validation was smaller. This will provide a convincing finding that in the equatorial region, the best performance for estimating the ZTD uses three input variables. Although the results of the performance evaluation model will increase the accuracy and the ability of ANFIS model in estimating the ZTD, a combination of the two input variables (P and H) and (P and T) is not suitable for the areas of Malaysia and Singapore. This can be seen from the Table 4.3, where the error generated is quite large and the correlation between estimated ZTD with ANFIS model and GPS ZTD is very weak. The results of second best performance in the estimation of ZTD value at three selected stations is used by two combination inputs of T and H, where its error is below the three input variables (P, T, and H). Table 4.3 shows that the configuration model with three inputs has smallest errors of below 0.1 %.

4.4.1   Comparison of ZTD from TroWav, IGS, and ANFIS

From processing of GPS data, we are now able to compare the ZTD values obtained from ANFIS FSC and ANFIS FCM models, GPS *TroWav*, and IGS. For Malaysia, we used the *TroWav* program to estimate the ZTD value. Indeed, the IGS ZTD at NTUS station is employed to compare the ZTD from *TroWav* and ANFIS models. The IGS ZTD data from NTUS is available with a resolution of five minutes.

Table 4.3 Statistical comparison between ZTD obtained from GPS and ANFIS models with four input networks at five selected stations in the Equatorial region during the process of training, testing, and validation. Both of RMSE and MAE are in the unit of meter

Site	Input	Training ANFIS FSC RMSE	MAE	ANFIS FCM RMSE	MAE	Testing ANFIS FSC RMSE	MAE	ANFIS FCM RMSE	MAE	Validation ANFIS FSC RMSE	MAE	ANFIS FCM RMSE	MAE
UKMB	P	0.022	0.018	0.022	0.018	0.022	0.017	0.021	0.017	0.020	0.016	0.020	0.016
	T	0.018	0.014	0.018	0.014	0.018	0.015	0.018	0.015	0.015	0.012	0.015	0.012
	H	0.018	0.014	0.018	0.014	0.018	0.014	0.018	0.014	0.016	0.013	0.016	0.013
	P, H	0.018	0.014	0.018	0.014	0.018	0.014	0.018	0.014	0.017	0.014	0.017	0.013
	P, T	0.015	0.012	0.017	0.014	0.017	0.014	0.017	0.014	0.017	0.014	0.015	0.012
	T, H	0.004	0.004	0.005	0.004	0.005	0.004	0.005	0.004	0.004	0.003	0.004	0.003
	P, T, H	0.000	0.000	0.000	0.000	0.000	0.000	0.000	0.000	0.000	0.000	0.000	0.000
NTUS	P	0.020	0.015	0.020	0.015	0.029	0.024	0.029	0.024	0.022	0.019	0.022	0.019
	T	0.016	0.013	0.016	0.013	0.025	0.021	0.024	0.021	0.019	0.016	0.019	0.016
	H	0.012	0.010	0.012	0.010	0.025	0.020	0.025	0.020	0.013	0.011	0.013	0.011
	P, H	0.012	0.009	0.011	0.009	0.025	0.019	0.025	0.019	0.014	0.011	0.014	0.011
	P, T	0.020	0.017	0.015	0.012	0.024	0.020	0.024	0.020	0.014	0.011	0.020	0.017
	T, H	0.008	0.006	0.008	0.006	0.004	0.003	0.004	0.003	0.004	0.004	0.004	0.004
	P, T, H	0.000	0.000	0.000	0.000	0.000	0.000	0.000	0.000	0.000	0.000	0.000	0.000
UMSK	P	0.016	0.013	0.016	0.013	0.014	0.011	0.014	0.011	0.020	0.016	0.020	0.016
	T	0.016	0.012	0.016	0.012	0.014	0.011	0.014	0.011	0.020	0.015	0.020	0.015
	H	0.013	0.010	0.013	0.010	0.012	0.010	0.012	0.010	0.016	0.012	0.016	0.012
	P, H	0.012	0.010	0.013	0.010	0.012	0.010	0.012	0.010	0.016	0.012	0.016	0.015
	P, T	0.016	0.012	0.015	0.012	0.013	0.010	0.013	0.011	0.020	0.015	0.019	0.015
	T, H	0.003	0.002	0.003	0.002	0.004	0.003	0.004	0.003	0.004	0.003	0.004	0.003
	P, T, H	0.000	0.000	0.000	0.000	0.000	0.000	0.000	0.000	0.000	0.000	0.000	0.000

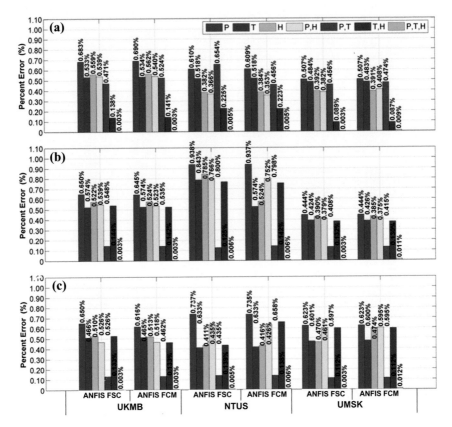

Fig. 4.24 Statistical comparison of error percentage (*PE*) between GPS ZTD and ANFIS ZTD models at the three selected stations in Equatorial region during the process of **a** training, **b** testing, and **c** validation

Figure 4.25 shows the comparison results of estimated ZTD for the case of December 2009 in a five-minute resolution.

The figure shows that there is a difference between ZTD generated from IGS and ZTD values generated from *TroWav* and ANFIS models, where the IGS ZTD is greater than the other methods. Moreover, the trend pattern of ZTD from *TroWav* and ANFIS models is not completely follows the trend of IGS ZTD. This comparison gives the impression that the IGS ZTD does not have a strong correlation with other methods. This condition shows that the ZTD from ANFIS models was trained using data from *TroWav* ZTD. Meanwhile, IGS ZTD is processed using GIPSY software with Niell hydrostatic mapping function with a cutoff elevation angle of 7° (Byun and Bar-Saver 2009). The *TroWav* ZTD is generated with VMF1 mapping function (Suparta 2014). On the other hand, a network configuration plan and design strategies in the processing the ZTD values may affect the product (Jin et al. 2007). Table 4.4 shows the statistical difference between IGS ZTD and

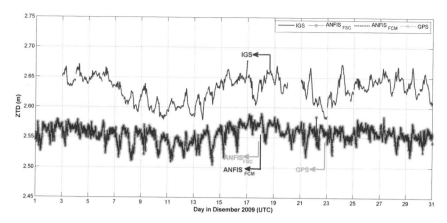

Fig. 4.25 Comparison results of estimated ZTD values through *TroWav* and ANFIS ZTD models with IGS ZTD at NTUS station for the case of December 2009

Table 4.4 Comparison of the difference method to estimate ZTD values between *TroWav* and ANFIS models with IGS ZTD at NTUS station for the case of December 2009. The ZTD values are expressed in cm with IGS ZTD being the larger

Difference	Minimum	Maximum	Average	RMS
$ZTD_{TroWav} - ZTD_{IGS}$	−1.392	−0.214	0.076	0.019
$ZTD_{ANFIS\ FSC} - ZTD_{IGS}$	−1.392	−0.214	0.076	0.019
$ZTD_{ANFIS\ FCM} - ZTD_{IGS}$	−1.391	−0.213	0.076	0.019

ZTD from *TroWav* and ANFIS models. From the Table 4.4, the range difference with the reference value of ZTD estimated using *TroWav* and ANFIS model is from −1.391 to −0.214 cm and the RMS is from 0.192 to 0.193 cm.

References

Bezdek JC, Ehrlich R, Full W (1984) FCM: the fuzzy c-means clustering algorithm. Comput Geosci 10(2):191–203

Boehm J, Werl B, Schuh H (2006) Troposphere mapping functions for GPS and very long baseline interferometry from European Centre for Medium-Range Weather Forecasts operational analysis data. J Geophys Res 111:B02406. doi:10.1029/2005JB003629

Byun SH, Bar-Sever YE (2009) A new type of troposphere zenith path delay product of the international GNSS service. J Geodesy 83(3–4):1–7

Chang FJ, Chang YT (2006) Adaptive neuro-fuzzy inference system for prediction of water level in reservoir. Adv Water Resour 29:1–10

Chaudhuri S, Middey A (2011) Adaptive neuro-fuzzy inference system to forecast peak gust speed during thunderstorms. Meteorol Atmos Phys 114:139–149. doi:10.1007/s00703-011-0158-4

Chiu SL (1994) Fuzzy model identification based on cluster estimation. J intell Fuzzy Syst 2(3):267–278

Dach R, Hugentobler U, Fridez, P, Meindl M (Eds) (2007) Astronomical Institute, University of Bern (AIUB): Bernese GPS Software Version 5.0

Gendt G (1998) IGS Combination of tropospheric estimates—experience from pilot experiment. In: Dow JM, Kouba J, Springer T (eds) Proc 1998 IGS Analysis Center Workshop. IGS Central Bureau, Jet Propulsion Laboratory, Pasadena, CA, pp 205–216

Gomez-Skarmeta AF, Delgado M, Vila MA (1999) About the use of fuzzy clustering techniques for fuzzy model identification. Fuzzy Set Syst 106(2):179–188

Hofmann-Wellenhof B, Lichtenegger H, Collins J (eds) (2001) Atmospheric effect on the global positioning system, theory and practice. Springer, Berlin

Jin S, Park JU, Cho JH, Park PH (2007) Seasonal variability of GPS-derived zenith tropospheric delay (1994–2006) and climate implication. J Geophys Res 112(D09110):1–11

Klobuchar JA, Parkinson BW, Spilker JJ (eds) (1996) Ionospheric effects in global positioning system: theory and applications. American Institute of Aeronautics and Astronautics, Washington DC

Klobuchar JA, Kunches JM (2003) Comparative range delay and variability of the earths troposphere and the ionosphere. GPS Solut 7:55–58. doi:10.1007/s10291-003- 0047-5

Lachapelle G (2003) Advanced GPS theory and applications. ENGO 625 Lecture Notes. Department of Geomatics Engineering, University of Calgary

Neshat M, Adeli A, Masoumi A, Sargolzae M (2011) A comparative study on ANFIS and fuzzy expert system models for concrete mix design. Int J Comput Sci Issues 8(3):196–210

Niell AE (1996) Global mapping functions for the atmosphere delay at radio wavelengths. J Geophys Res 101(B2):3197–3246

Nourani V, Komasi M (2013) A geomorphology-based ANFIS model for multi-station modeling of rainfall–runoff process. J Hydrol 490:41–55

Park J, Kim JW, Chang JH, Jin YG, Kim NS (2015) Estimation of speech absence uncertainty based on multiple linear regression analysis for speech enhancement. Appl Acoust 87:205–211. doi:10.1016/j.apacoust.2014.06.017

Suparta W (2014) The development of GPS TroWav tool for atmospheric—terrestrial studies. J Phys Conf Ser 495:012037. doi:10.1088/1742-6596/495/1/012037

Suparta W, Alhasa KM (2013) A comparison of ANFIS and MLP models for the prediction of precipitable water vapor. In: Proceedings of 2013 3rd IEEE international conference on space science and communication (IconSpace2013), pp 243–247. doi:10.1109/IconSpace.2013. 6599473

Suparta W, Alhasa KM (2015) Modeling of zenith path delay over Antarctica using an adaptive neuro fuzzy inference system technique. Expert Syst Appl 42(3):1050–1064

Suparta W, Mandeep Singh JS, Mohd Alauddin MA, Yatim B, Mohd Yatim AN (2011) GPS water vapor monitoring and TroWav updated for ENSO studies. In: The 2nd international conference on instrumentation, communications, information technology, and biomedical engineering (ICICI-BME) 2011, pp 35–39

Yager RR, Filev DP (1994) Generation of fuzzy rules by mountain clustering. J Intell Fuzzy Syst 2 (3):209–219

Yu H, Wilamowski BM (2011) Levenberg-marquardt training. The Industrial Electronics Handbook, pp 1–15

Chapter 5
Prediction of ZTD Based on ANFIS Model

Abstract Since the tropospheric delay plays a crucial role in meteorological studies and weather forecasts as well as positioning accuracy, accurate prediction of its value is critical to helping monitoring the ZTD variation on a global basis. In the previous chapter, ZTD has been estimated with a fuzzy inference system that uses a back-propagation algorithm. The input of the system is surface meteorological data and the test output is ZTD from GPS. For a test case, a combination of surface pressure (P), temperature (T), or relative humidity (H) is performed to obtain the best estimation of ZTD model. Based on the prospect of ZTD estimation using ANFIS, this chapter will focus on how to predict ZTD value using the surface meteorological data.

Keywords Estimation of ZTD · ANFIS models · Configuration inputs · Surface meteorological data · Predictive model

5.1 Configuration Model for Estimation and Prediction of ZTD

After successfully applying the ANFIS models to estimate and investigate the ZTD value from three inputs of surface meteorological data, we developed the ANFIS model to predict ZTD value. We determine the predictive model from one to a certain step ahead to measure how high the accuracy of ZTD predicted from the previous data availability. Table 5.1 shows the two different groups (I, II) of model configuration input used for prediction of ZTD. We developed one input combination with a different number of variables (P, T, and H) and inserted into the input layer to predict the ZTD value. The best structure of ANFIS model was examined according to the evaluation criteria. Afterward, the selection of input variables and determining the architecture of ANFIS models such as type and number of membership functions (MFs) are considered to satisfy the predictive models developed.

© The Author(s) 2016
W. Suparta and K.M. Alhasa, *Modeling of Tropospheric Delays Using ANFIS*,
SpringerBriefs in Meteorology, DOI 10.1007/978-3-319-28437-8_5

Table 5.1 Summary of configuration inputs used in the development of ANFIS model for estimation and prediction of ZTD

Group	Combination	Input	Output
I	A1	$P(t)$	ZTD(t)
	A2	$T(t)$	ZTD(t)
	A3	$H(t)$	ZTD(t)
	A4	$P(t)$, $H(t)$	ZTD(t)
	A5	$P(t)$, $T(t)$	ZTD(t)
	A6	$T(t)$, $H(t)$	ZTD(t)
	A7	$P(t)$, $T(t)$, $H(t)$	ZTD(t)
II	B1	$P(t - 1\ s)$, $P(t)$, $T(t - 1\ s)$, $T(t)$, $H(t - 1\ s)$, $H(t)$	ZTD$(t + i)$

This includes the selection of parameters in the training procedure, such as the epoch number, training goal, and step sizes. Figure 5.1 summarizes how to get the ZTD prediction using ANFIS model (Suparta and Alhasa 2013a).

5.2 Prediction Results

5.2.1 Results from Selected Stations in Antarctica

After the training process, the ANFIS models showed that they were able to capture all the characteristics of ZTD. Since the meteorological data obtained was missing especially for OHI2 stations, we used the November data for the validation process due to the data accessibility at the five selected stations. Figure 5.2 showed the results of the general performance of the constructed ZTD model from one-step to eight-step ahead for both models (FCM (I) and FSC (II)) in comparison with GPS ZTD at selected stations in Antarctica. The bottom graph in each panel shows the ZTD data from GPS ZTD. The graph above GPS ZTD as indicated by $t + 1$ to $t + 8$ is the prediction result with meteorological data as input.

As demonstrated in Fig. 5.2, the blank value in the figure shows no data, either GPS or meteorological data, and the ZTD value in each step obtained from prediction results is shifted every 0.10 m from GPS ZTD to see more clearly their variations. The similar prediction result for selected stations in Antarctica Peninsula is presented in Fig. 5.3. As shown in both figures, two ANFIS models (ANFIS FSC and ANFIS FCM) were used to build the predictive models at all stations. Based on data availability, we use data at intervals of 10-min (SBA), 5-min (DAV1), and 3-h (SYOG, PALM, and OHI2) in the development of predictive models for selected stations in Antarctica. The results clearly show that the models were able to capture all the characteristics that existed in the GPS ZTD value, and could predict ZTD very well from one-step to eight-step ahead. On the right panel of each figure also indicated prediction step from $t + 1$ until $t + 8$, which means that $t + 1$ is ZTD

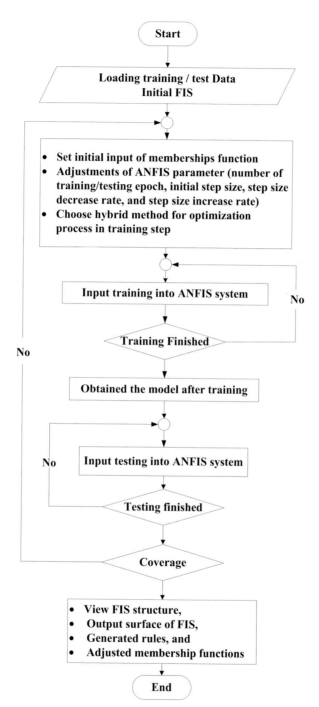

Fig. 5.1 Flowchart of prediction of ZTD from ANFIS model

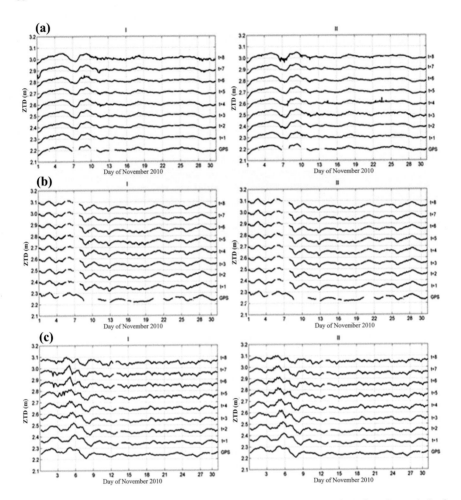

Fig. 5.2 Prediction of ZTD trend from one-step to eight-step ahead during the period of November 2010 for Antarctica Coast at **a** SBA, **b** DAV1, and **c** SYOG stations for ANFIS FCM (*I*) and ANFIS FSC (*II*), respectively

prediction for the first 10 min (SBA), or for the first 3-h (SYOG, PALM, and OHI2). For $t + 5$, it means that the ZTD prediction is in the 50th min (SBA), or prediction for SYOG, PALM, and OHI2 at 15:00 h, while ZTD prediction at $t + 0$ is obtained similar with the value of GPS ZTD. In other words, the figure shows the time step of prediction which depends on the interval data used.

Figure 5.4 summarizes the entire value of estimation errors or accuracy of ZTD prediction for Antarctica. The error was found to be below 1 %, where (a) SBA and (b) DAV1 showed an almost similar trend of estimation errors compared to (c) SYOG, (d) PALM, and (e) OHI2. The trend of estimation errors for the last three stations with a 3-h interval will increase when each step ahead of prediction is

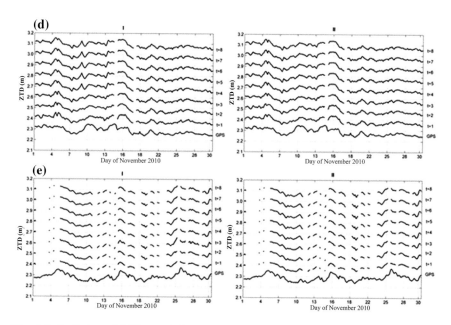

Fig. 5.3 Prediction of ZTD trend from one-step to eight-step ahead during the period of November 2010 for Antarctica Peninsula: **d** PALM and **e** OHI2 stations

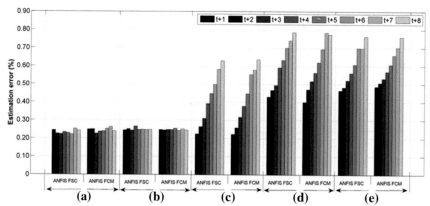

Fig. 5.4 The estimation errors of ZTD prediction from one-step until eight-step ahead using the ANFIS models compared to GPS ZTD for **a** SBA, **b** DAV1, **c** SYOG, **d** PALM, and **e** OHI2 stations

increased. This increasing trend confirmed due to a low data interval (3-h) used as compared to other two stations. The prediction can be made for more than $t + 8$, however, its accuracy will be reduced significantly.

Table 5.2 showed the relationship of ZTD obtained between GPS and ANFIS models (FSC and FCM) from one-step until eight-step ahead. The table simplified

Table 5.2 Correlation of determination (R^2) for prediction of ZTD obtained between GPS and ANFIS models (FSC and FCM) from one-step until eight-step ahead at five selected stations in Antarctica

Step	SBA		DAV1		SYOG		PALM		OHI2	
	FSC	FCM	FSC	FCM	FSC	FCM	FSC	FCM	FSC	FCM
$t + 1$	0.961	0.956	0.940	0.939	0.949	0.950	0.825	0.824	0.789	0.759
$t + 2$	0.956	0.964	0.938	0.940	0.926	0.926	0.790	0.785	0.773	0.759
$t + 3$	0.958	0.958	0.941	0.942	0.885	0.879	0.737	0.756	0.733	0.729
$t + 4$	0.950	0.942	0.912	0.942	0.819	0.824	0.680	0.637	0.690	0.689
$t + 5$	0.965	0.959	0.940	0.936	0.747	0.751	0.622	0.604	0.642	0.644
$t + 6$	0.963	0.957	0.942	0.942	0.679	0.629	0.520	0.495	0.564	0.594
$t + 7$	0.960	0.940	0.942	0.924	0.587	0.569	0.431	0.465	0.511	0.537
$t + 8$	0.957	0.948	0.941	0.941	0.516	0.483	0.428	0.423	0.472	0.485

that the coefficient of determination (R^2) value obtained was larger than 0.90, which clarified that the ANFIS models are accurate and consistent in different subsets. However, at seven and eight steps predictions of SYOG, the ANFIS models were not able to cover all the characteristics of ZTD value, as indicated by its lower R^2 values (below 0.60). The (d) PALM and (e) OHI2 stations with a 3-h data resolution had a similar trend of estimation error with (c) SYOG, where the trend error will be gradually increased when the step ahead increased. In contrast to other stations in the Antarctica coast, (d) PALM, and (e) OHI2 were observed to have R^2 values lower than 0.80. The moderate values of R^2 in these two stations for increasing step prediction ahead were possibly due to the atmospheric conditions in the Antarctica peninsula that had affected the model accuracy. Overall, the performance of the two ANFIS models developed was very good as indicated by smaller estimation error (see Fig. 5.4), which clarifies that the ANFIS models can be an alternative method to predict the ZTD value (Suparta and Alhasa 2015). From Figs. 5.3, 5.4, and Table 5.2, we found that one-step until eight-step ahead was the best in a predictive model for Antarctica region based on the data resolution used. On the other hand, the prediction with more than nine steps will significantly decrease its accuracy. We noted that the accuracy of the predictive model developed, in this case, will be dependent on the data resolution available.

5.2.2 Results from Selected Stations in Equatorial Region

For prediction of ZTD value at selected stations in Equatorial region, the same structure and algorithm as in Antarctica were employed. For the prediction development, we used a one-min interval of surface meteorological data (P, T, and H) for all stations selected as inputs. To realize the configuration in group II ($B1$), the overall data for November 2009 has been chosen for UKM station, and month of December 2012 was chosen for NTUS and UMSK stations. Figure 5.5 shows the

(a)

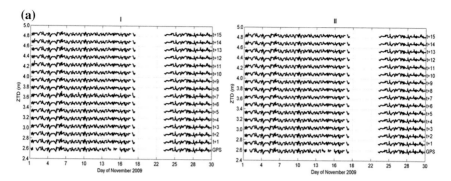

Fig. 5.5 Prediction of ZTD trend from one-step to fifteen-step ahead during the period of November 2009 for **a** UKMB station in equatorial region

prediction of ZTD results for UKMB station. With 1-min data resolution, the prediction can be made until $t + 15$ for the best ZTD value. As shown in the figure, the blank value in the figure shows no data for the period from 17 to 23 November 2009 either GPS or meteorological data. In addition to the similar presentation results as in Antarctica, the ZTD value in each step is shifted every 0.10 m from GPS ZTD value.

Similar figure, as in Fig. 5.5 for NTUS and UMSK stations, is depicted in Fig. 5.6. The blank data for UMSK station was recorded for the period from 18 to 29 December 2012. The trend and variation of predicted ZTD showed no difference from GPS ZTD data, and for this result, the prediction from one-step to fifteen-step ahead obtained with a very good correlation. Figure 5.7 summarizes the entire value of estimation errors in ZTD prediction for selected stations in Equatorial region. The error was found to be below 0.18 %, where (a) UKMB and (c) UMSK showed an almost similar trend of estimation errors compared to (b) NTUS. The trend of estimation errors for (b) NTUS is in the range of 0.07 % to below 0.16 % with 0.01 % different between ANFIS FSC and ANFIS FCM for all steps prediction. The low error for NTUS is due to no data missing as compared to the other two stations. Table 5.3 compiled the relationship of ZTD obtained between GPS and ANFIS models (FSC and FCM) from one-step until fifteen-step ahead. The table simplified that the R^2 value obtained was larger than 0.90, which clarified that the ANFIS models are accurate and consistent in different subsets.

The combination $B1$ (group II) as the input layer with a gauss function was selected as a membership function to satisfy the structure of ANFIS models. We noted that the number of rules should be determined carefully to prevent over-parameterization of ANFIS model and to improve time efficiency requested for training structure and determining parameters. In addition, the selection of parameters such as training step and iteration number is important because the appropriate number can progress the model efficiency in training, test, and validation steps, and prevents the model to be over trained. After training, the different structure of ANFIS model via different epoch number and step size parameters

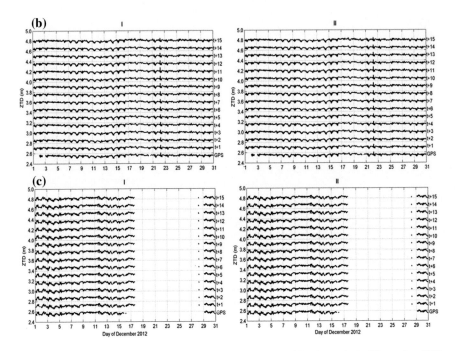

Fig. 5.6 Prediction of ZTD from one-step to fifteen-step ahead for the period of December 2012 at **b** NTUS, and **c** UMSK stations in for equatorial region

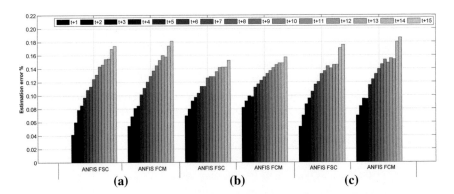

Fig. 5.7 The estimation errors of ZTD prediction from one-step until fifth teen-step ahead using the ANFIS models compared to GPS ZTD for **a** UKMB, **b** NTUS, and **c** UMSK stations

found that 200 epochs satisfied the training network with 2×10^{-4} as the goal of performance and 0.05 as the initial step size. With this network, the results demonstrated that ANFIS has been successfully applied to establish the predictive models that could provide accurate and reliable ZTD values.

Step	UKMB		NTUS		UMSK	
	FSC	FCM	FSC	FCM	FSC	FCM
$t+1$	0.980	0.978	0.954	0.949	0.989	0.984
$t+2$	0.963	0.961	0.945	0.938	0.981	0.976
$t+3$	0.949	0.946	0.933	0.929	0.973	0.969
$t+4$	0.938	0.934	0.926	0.925	0.967	0.967
$t+5$	0.926	0.921	0.918	0.911	0.960	0.956
$t+6$	0.913	0.909	0.907	0.904	0.953	0.949
$t+7$	0.908	0.898	0.904	0.897	0.949	0.943
$t+8$	0.894	0.887	0.891	0.890	0.941	0.937
$t+8$	0.894	0.887	0.891	0.890	0.941	0.937
$t+9$	0.885	0.876	0.885	0.883	0.936	0.931
$t+10$	0.868	0.866	0.883	0.876	0.930	0.925
$t+11$	0.868	0.856	0.873	0.869	0.933	0.929
$t+12$	0.847	0.846	0.864	0.861	0.928	0.923
$t+13$	0.830	0.829	0.862	0.858	0.928	0.924
$t+14$	0.828	0.826	0.857	0.848	0.905	0.899
$t+15$	0.828	0.817	0.844	0.842	0.901	0.893

Table 5.3 Correlation of determination (R^2) for prediction of ZTD obtained between GPS and ANFIS models (FSC and FCM) from one-step until fifteen-step ahead at three selected stations in equatorial region

5.3 Estimation of ZTD Value from Model Developed

Group I ($A_1 \ldots A_7$) from Table 5.1 is set for estimation of ZTD value with an input of surface meteorological data. We estimated the ZTD for three selected stations (CAS1, MCM4, and SYOG) in Antarctica. Since the other input or its combination was shown weak correlation, we only selected four inputs with (a) P, (b) P and T, (c) P and H, and (d) P, T, and H. In this test case, ZTD value is estimated by using both ANFIS FCM and ANFIS FSC models. The curve used in the formation of the membership function is 'gaussmf'. The formation rule of Sugeno-Takagi with FIS type is using a linear equation. There are two linear equations in this FIS that has been optimized to estimate ZTD as expressed in Eqs. 5.1 and 5.2.

$$\text{ZTD ANFIS FSC} = \sum_{i=1,2}^{2} \overline{w_i} f_i = \overline{w_i}(A_i(P) + B_i(T) + C_i(H) + D_i) \qquad (5.1)$$

$$\text{ZTD ANFIS FCM} = \sum_{i=1,2}^{2} \overline{w_i} f_i = \overline{w_i}(A_i(P) + B_i(T) + C_i(H) + D_i) \qquad (5.2)$$

where $\overline{w_{i=1,2}}$ is taken as the normalized firing strength in the third layer and ZTD is in meter. If the predicate α for the two rules are w_1 and w_2, then the weighted average for ZTD can be calculated as

$$\text{ZTD} = \frac{w_1 f_1 + w_2 f_2}{w_1 + w_2} = \overline{w_1} f_1 + \overline{w_2} f_2 \qquad (5.3)$$

$\overline{w_1}$ and $\overline{w_2}$ can be calculated using Eqs. 2.5–2.8 (see Chap. 2), where Eq. 2.4 is calculated using '*gaussmf*' function in MATLAB. The rule values (consequent parameters), f_1 and f_2 are obtained by a least square method. Now, with the Eq. 5.3, the ZTD at each station can be determined either using FSC or FCM model. For each station, the linear equation for FSC and FCM are obtained as follows:

CAS1:

$$f_{1,\text{FSC}} = 0.002067P + 0.001170T - 0.000158200H + 0.25440$$
$$f_{2,\text{FSC}} = 0.001775P + 0.001090T + 0.00009403H + 0.54720$$
$$f_{1,\text{FCM}} = 0.002237P + 0.001783T + 0.00004162H + 0.08865$$
$$f_{2,\text{FCM}} = 0.002236P + 0.002386T + 0.00042290H + 0.05803$$

MCM4:

$$f_{1,\text{FSC}} = 0.002150P + 0.001103T - 0.00001739H + 0.11540$$
$$f_{2,\text{FSC}} = 0.002339P + 0.001541T + 0.00038910H - 0.09675$$
$$f_{1,\text{FCM}} = 0.002106P + 0.000939T - 0.0001247H + 0.14760$$
$$f_{2,\text{FCM}} = 0.002313P + 0.001307T - 0.0005171H - 0.08295$$

SYOG:

$$f_{1,\text{FSC}} = 0.001985P + 0.001833T + 0.00009787H + 0.3283$$
$$f_{2,\text{FSC}} = 0.002286P + 0.001246T + 0.00021780H + 0.0098$$
$$f_{1,\text{FCM}} = 0.002202P + 0.001210T + 0.00009170H + 0.0990$$
$$f_{2,\text{FCM}} = 0.002202P + 0.003363T - 0.00040200H + 0.1631$$

where P is the surface pressure (mbar), T is the surface temperature (in degree Celsius), and H is the relative humidity (in percent). After w_1 and w_2 obtained, then ZTD value can be computed using Eq. 5.3.

Figure 5.8 shows the ZTD estimation for Casey station (CAS1). A similar result for MCM4 and SYOG is presented in Figs. 5.9 and 5.10, respectively. From the selected surface meteorological data as an input, ZTD estimated with three inputs of surface meteorological data shows the best performance with high accuracy compared with only one input P, T, or H. Alternatively, the model with two inputs such as pressure and temperature (P, T) will also be competitive in estimating the ZTD. However, it has been tested that estimation of ZTD without P, or just with T or H alone or a combination of T and H as input, will not get any correspondence with the trend of GPS ZTD. This is to confirm that the combination of T and H,

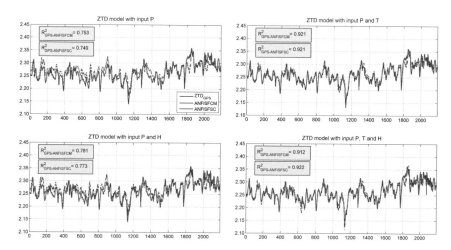

Fig. 5.8 Variation of ZTD estimated from ANFIS FCM and ANFIS FSC models compared with ZTD data from GPS at CAS1 station, Antarctica for the period from July to December 2005, after validation process

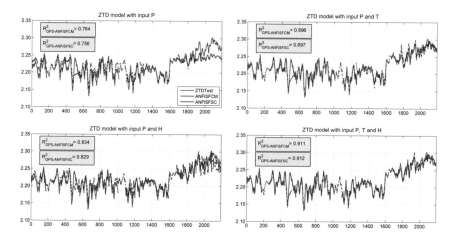

Fig. 5.9 Variation of ZTD estimated from ANFIS FCM and ANFIS FSC models compared with ZTD data from GPS at MCM4 station, Antarctica for the period from July to December 2005, after the validation process

or T or H alone as an input in Table 5.1 is not employed to estimate ZTD. For Antarctica region, the surface pressure likely plays a major role in determining the characteristic of the atmosphere at the target station. The statistical result of model comparison from a given input to estimate ZTD from the two ANFIS models is presented in Table 5.4. From the table, it is shown that ANFIS FSC model is almost comparable with ANFIS FCM model for estimation of ZTD.

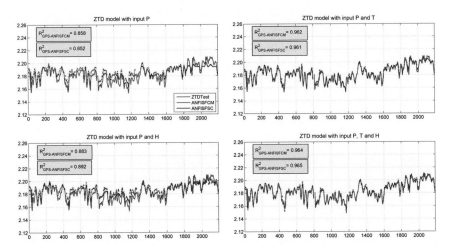

Fig. 5.10 Variation of ZTD estimated from ANFIS FCM and ANFIS FSC models compared with ZTD data from GPS at SYOG station, Antarctica for the period from July to December 2005, after the validation process

Table 5.4 Statistical result to estimate ZTD between the two models compared to GPS ZTD for the period from July to December 2005 at three selected stations over Antarctica. The unit for MAE and RMSE is in cm

Model	Input	Casey (CAS1)			McMurdo (MCM4)			Syowa (SYOG)		
		MAE	PE	RMSE	MAE	PE	RMSE	MAE	PE	RMSE
ANFIS FSC	P	1.52	0.67	1.85	1.42	0.64	1.81	1.37	0.61	1.59
	P, H	1.41	0.63	1.73	1.15	0.52	1.49	1.29	0.57	1.52
	P, T	0.73	0.32	0.96	0.74	0.33	1.10	0.45	0.20	0.60
	P, T, H	0.72	0.31	0.96	0.65	0.29	1.01	0.43	0.19	0.57
ANFIS FCM	P	1.52	0.67	1.84	1.38	0.62	1.77	1.36	0.60	1.59
	P, H	1.44	0.63	1.75	1.16	0.52	1.50	1.30	0.58	1.52
	P, T	0.73	0.32	0.96	0.74	0.33	1.10	0.44	0.20	0.59
	P, T, H	0.79	0.35	1.04	0.66	0.29	1.02	0.44	0.19	0.59

Figure 5.11 shows the example of ZTD estimation with three meteorological data (P, T, and H) as the input for a one-hour interval. As shown in the figure, the variation of ZTD from both models is very closer. In this case, the ANFIS FSC is more accurate to 2.3 % compared to the ANFIS FCM method. In addition, estimation of ZTD using ANFIS techniques with three surface meteorological data as the inputs are more promising with an accuracy of 20 % higher compared to other inputs. This would provide a new alternative where the GPS data is not necessary to determine ZTD (Suparta and Alhasa 2013b).

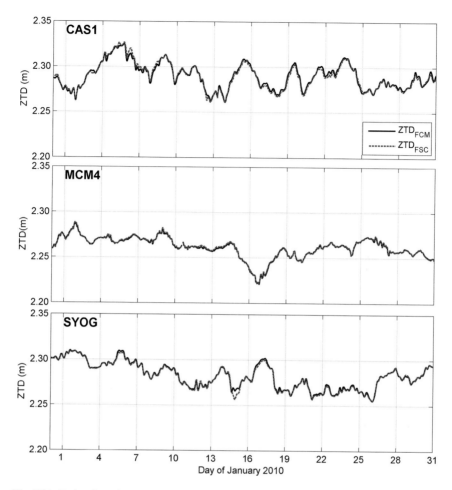

Fig. 5.11 Estimation of ZTD from three surface meteorological data at three selected stations in Antarctica for the period of January 2010

5.4   Prediction of ZTD Value from Model Developed

After successfully estimating ZTD value using group I configuration, and based on the accuracy demonstrated in Figs. 5.4 and 5.7, and the strong correlation between ANFIS ZTD and GPS ZTD in Tables 5.2 and 5.3, we employed the surface meteorological data for prediction of ZTD. This prediction is proposed to anticipate the unavailable future data for the estimation process. For the selected surface meteorological data as an input, the combination input layer B1 (group II) was used to predict ZTD value at selected stations. We use both the models (FCM and FSC) that have been trained to predict the ZTD value.

The equation to calculate of ZTD prediction is expressed in Eq. 5.4. This equation is similar to Eq. 5.3, which is capable to determine the ZTD prediction at each station either using ANFIS FSC or ANFIS FCM models.

$$ZTD(t+i) = \frac{w_1 f_1 + w_2 f_2 + \cdots + w_n f_n}{w_1 + w_2 + \cdots + w_n}, \quad i = n = 1, 2, 3, \ldots m \quad (5.4)$$

where w_1, w_2, and w_n is obtained from multiple incoming signals membership grades of the input variable as shown in Eq. 5.5, t is time, and f_1, f_2, and f_n are rules (consequent parameters) or linear equations for the output obtained by a least square method.

$$w_i = \mu_{P(t-1)i} \cdot \mu_{P(t)i} \cdot \mu_{T(t-1)i} \cdot \mu_{T(t)i} \cdot \mu_{H(t-1)i} \cdot \mu_{H(t)i}, \quad i = 1, 2, 3 \ldots n \quad (5.5)$$

where μ is a membership grade of input variable (P, T, and H) calculated using the gauss membership function equation (see group II or B1 configuration), as expressed in Eq. 5.6. On the other hand, w_i represents the firing strength of a rule.

$$\mu = \exp\left[-\frac{1}{2}\left(\frac{x - c_i}{a_i} \right)^2 \right], \quad i = 1, 2, 3 \ldots n \quad (5.6)$$

where a (sigma) and c (center) are called $\{a, c\}$ as premise parameters and x is a data point (e.g., P, T, or H).

For the assessment in Antarctica region, we selected surface meteorological data on 1–2 January 2011. For an example case, we predict ZTD for SYOG station at $t + 4$ ($i = 4$ from Eq. 5.4). Note that the surface meteorological data at SYOG station is available at a 3-h of data resolution. By this resolution, we present two days of data (48 h) to see the trend of prediction results. Although the data is more than one day, the ZTD is predicted by day per day and superimposed in a graph. The premise and consequent parameters for the four steps ahead that have been trained for ANFIS FCM and ANFIS FSC models are compiled in Tables 5.5, 5.6, 5.7, and 5.8, respectively. Each input is to have a gauss membership function $\{a, c\}$. From Tables 5.7 and 5.8, we found seven rules based on the trial-and-error process to obtain an optimum value of ZTD prediction. The prediction of ZTD is referring to the value and trend of GPS ZTD from IGS. Then in each rule, we can produce seven equations, where each equation is with three inputs (P, T, and H).

From Tables 5.5 and 5.6, the ZTD can be computed using Eq. 5.4 where consequent parameter (f) in each rule in Table 5.6 is summed (from f_1 to f_7) and the value of $\{a, c\}$ (membership functions) in Table 5.5 is computed using Eq. 5.5. The coefficient or value in front of each variable input in Table 5.6 (e.g., 5.424×10^{-292} in the first rule) is obtained using a least squared method from ANFIS Toolbox of MATLAB. The number of rules is obtained based on the trial-and-error process. A similar method for prediction of ZTD based on the ANFIS FSC at $t + 4$ ($\sim 12{:}00$ h) for SYOG is presented in Tables 5.7 and 5.8. It is clearly shown that

Table 5.5 The premise of ANFIS FCM model for $t + 4$ at SYOG station

Rule	if—part[a]											
	$P(t − 1\ s)$		$P(t)$		$T(t − 1\ s)$		$T(t)$		$H(t − 1\ s)$		$H(t)$	
	a	c	a	c	a	c	a	c	a	c	a	c
1	54.97	973	50.68	985.0	21.78	−24.98	−44.05	12.36	35.50	60.05	32.40	55.45
2	−51.14	1014	−36.88	959.6	−1.519	19.48	24.29	32.84	29.11	150.3	68.83	79.65
3	2.316	1035	13.82	1002	21.78	24.98	50.74	33.55	47.74	61.67	−61.93	84.12
4	42.20	1008	−4.406	1044	36.84	−1.268	35.81	11.23	71.12	59.56	81.30	43.00
5	−42.90	985	−32.65	997.3	−152.7	−148.3	47.43	37.41	−56.8	58.01	−67.12	80.48
6	−12.30	997.1	164.7	1024	20.12	21.19	−139.3	10.80	32.60	56.32	30.60	54.39
7	11.60	992.2	47.57	979.1	−1.519	19.48	9.853	10.25	142	153.0	40.55	143.5

[a]$\{a, c\}$ are gauss membership function obtained using Eq. 5.4

Table 5.6 The consequence parameter of ANFIS FCM model for $t + 4$ at SYOG station

Rule	Then—part
	Equation for f = consequent parameter
1	$3.971 \times 10^{-3}P(t - 1 \text{ s}) + 5.369 \times 10^{-3}P(t) + 6.893 \times 10^{-4}T(t - 1 \text{ s}) + -1.522 \times 10^{-3}$ $T(t) + -4.839 \times 10^{-4}H(t - 1 \text{ s}) + 4.466 \times 10^{-4}H(t) + -0.248$
2	$6.583 \times 10^{-3}P(t - 1 \text{ s}) + -7.503 \times 10^{-3}P(t) + -0.138T(t - 1 \text{ s}) + 0.215$ $T(t) + 1.354 \times 10^{-2}H(t - 1 \text{ s}) + 4.021 \times 10^{-2}H(t) + 0.646$
3	$4.295 \times 10^{-22}P(t - 1 \text{ s}) + 4.290 \times 10^{-22}P(t) + -4.333 \times 10^{-25}$ $T(t - 1 \text{ s}) + -5.090 \times 10^{-25}T(t) + 3.682 \times 10^{-23}H(t - 1 \text{ s}) + 3.664 \times 10^{-23}$ $H(t) + 4.261 \times 10^{-25}$
4	$4.103 \times 10^{-8}P(t - 1 \text{ s}) + 4.104 \times 10^{-8}P(t) + 1.022 \times 10^{-11}T(t - 1 \text{ s}) + 1.670 \times 10^{-11}$ $T(t) + 3.608 \times 10^{-9}H(t - 1 \text{ s}) + 3.590 \times 10^{-9}H(t) + 4.072 \times 10^{-11}$
5	$-2.868 \times 10^{-3}P(t - 1 \text{ s}) + 3.198 \times 10^{-3}P(t) + -6.403 \times 10^{-4}T(t - 1 \text{ s}) + 7.534 \times 10^{-3}$ $T(t) + 1.280 \times 10^{-3}H(t - 1 \text{ s}) + 1.864 \times 10^{-3}H(t) + 2.081$
6	$-7.981 \times 10^{-3}P(t - 1 \text{ s}) + 9.620 \times 10^{-3}P(t) + -2.162 \times 10^{-3}T(t - 1 \text{ s}) + 3.225 \times 10^{-3}$ $T(t) + 2.909 \times 10^{-4}H(t - 1 \text{ s}) + 9.033 \times 10^{-4}H(t) + 0.623$
7	$-6.59 \times 10^{-28}P(t - 1 \text{ s}) + -6.566 \times 10^{-28}P(t) + -3.104 \times 10^{-31}$ $T(t - 1 \text{ s}) + 6.115 \times 10^{-32}T(t) + -5.800 \times 10^{-29}H(t - 1 \text{ s}) + -5.761 \times 10^{-29}$ $H(t) + -6.582 \times 10^{-31}$

the value of membership functions $\{a, c\}$ between the two table is not similar although it is carried out with the same process. The different result is due to the different method used (FCM or FSC) to generate FIS. By a different initial, FIS will result in different values in training.

Figure 5.12 shows the result of ZTD prediction for $t + 4$ or at 12 h ahead at SYOG station. The figure showed the ZTD predicted by ANFIS (FCM and FSC) models and ZTD from GPS is plotted for 48 h. ZTD from GPS in this tested case was obtained from the database of Crustal Dynamics Data Information System (CDDIS) NASA (see Sect. 4.2.2) with the resolution of one-min. As shown in the figure, ZTD trend from ANFIS models at $t + 4$ are closely correlated and are exhibited with an almost similar trend to the GPS ZTD. However, the ANFIS ZTD is only captured about 78 % compared to ZTD trend from GPS. This indicated that more steps of prediction will decrease its trend accuracy. Data resolution also plays a crucial role in determining the accuracy of prediction. In addition, more step of prediction will lead to more decrease of its accuracy.

For the assessment of Equatorial region, we selected surface meteorological data on January 1st, 2010. For an example case, we predict ZTD for $t + 8$ for NTUS station. The surface meteorological data (P, T, and H) for ZTD prediction is with a one-min interval. We use a similar approach as in Antarctica, where the premise and consequent parameters for eight steps ahead that has been trained are compiled in Tables 5.9 and 5.10 (ANFIS FCM) and Tables 5.11 and 5.12 (ANFIS FSC). In the ANFIS FCM model, we found three rules to obtain the best prediction of ZTD. Then it has been chosen when the ANFIS model reaches an acceptable satisfactory level. Note that the ZTD from GPS for NTUS is computed using *TroWav* program (Suparta et al. 2008; Suparta 2014).

Table 5.7 The premise of ANFIS FSC model for $t + 4$ at SYOG station

Rule	if—part[a] P(t − 1 s)		P(t)		T(t − 1 s)		T(t)		H(t − 1 s)		H(t)	
	a	c	a	c	a	c	a	c	a	c	a	C
1	26.45	1009	0.265	1018	−105.7	3.189	−54.0	2.806	−51.3	78.84	42.13	96.00
2	44.09	941.1	−50.89	915.4	−15.94	35.23	81.23	−46.2	59.96	102.0	65.61	82.35
3	39.33	945.4	−42.80	983.2	70.14	−0.614	64.01	−17.0	−74.3	25.85	46.37	44.87
4	59.35	993.2	17.06	1002	−28.70	20.37	20.51	17.25	30.29	85.79	11.70	79.70
5	14.16	950.6	21.94	966.9	37.67	−22.86	21.48	−24.3	39.94	91.16	54.11	74.53
6	−42.80	1025	28.59	978.5	49.70	22.90	21.00	44.20	63.02	57.91	62.42	104.9
7	56.29	975.6	55.53	975.1	39.33	5.742	6.249	43.59	77.30	35.38	69.37	77.33

[a]{a, c} are gauss membership function obtained using Eq. 5.4

Table 5.8 The consequence parameter of ANFIS FSC model for $t + 4$ at SYOG station

Rule	Then—part
	Equation for f = consequent parameter
1	$5.424 \times 10^{-292} P(t - 1 \text{ s}) + 5.425 \times 10^{-292} P(t) + -3.767 \times 10^{-295}$ $T(t - 1 \text{ s}) + -5.920 \times 10^{-295} T(t) + 4.706 \times 10^{-293} H(t - 1 \text{ s}) + 4.425 \times 10^{-293}$ $H(t) + 5.382 \times 10^{-295}$
2	$-4.906 \times 10^{-3} P(t - 1 \text{ s}) + 9.469 \times 10^{-3} P(t) + 4.146 \times 10^{-4} T(t - 1 \text{ s}) + -1.324 \times 10^{-3}$ $T(t) + -4.716 \times 10^{-4} H(t - 1 \text{ s}) + 5.994 \times 10^{-4} H(t) + -2.156$
3	$-4.934 \times 10^{-3} P(t - 1 \text{ s}) + 6.387 \times 10^{-3} P(t) + -3.809 \times 10^{-4} T(t - 1 \text{ s}) + 2.048 \times 10^{-3}$ $T(t) + -1.536 \times 10^{-4} H(t - 1 \text{ s}) + 2.223 \times 10^{-4} H(t) + 0.871$
4	$-1.608 \times 10^{-2} P(t - 1 \text{ s}) + 1.860 \times 10^{-2} P(t) + -7.392 \times 10^{-3} T(t - 1 \text{ s}) + 1.123 \times 10^{-2}$ $T(t) + -7.124 \times 10^{-4} H(t - 1 \text{ s}) + 2.224 \times 10^{-3} H(t) + 0.0320$
5	$-1.196 \times 10^{-3} P(t - 1 \text{ s}) + 7.315 \times 10^{-4} P(t) + -2.730 \times 10^{-3} T(t - 1 \text{ s}) + 1.761 \times 10^{-3}$ $T(t) + 4.747 \times 10^{-4} H(t - 1 \text{ s}) + -1.790 \times 10^{-4} H(t) + 2.626$
6	$6.310 \times 10^{-2} P(t - 1 \text{ s}) + -6.587 \times 10^{-2} P(t) + 0.014 T(t - 1 \text{ s}) + -1.034 \times 10^{-2}$ $T(t) + 9.761 \times 10^{-3} H(t - 1 \text{ s}) + 8.589 \times 10^{-3} H(t) + 3.763$
7	$-6.859 \times 10^{-4} P(t - 1 \text{ s}) + -6.750 \times 10^{-4} P(t) + -1.324 \times 10^{-6} T$ $(t - 1 \text{ s}) + -1.147 \times 10^{-6} T(t) + -5.464 \times 10^{-5} H(t - 1 \text{ s}) + -5.149 \times 10^{-5}$ $H(t) + -6.519 \times 10^{-7}$

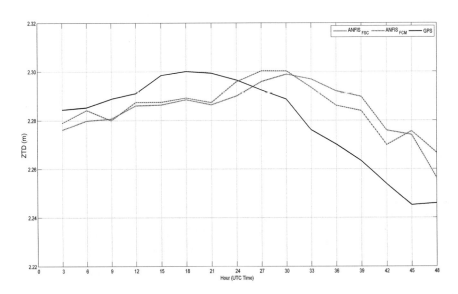

Fig. 5.12 The prediction of ZTD for SYOG station at $t + 4$ for the period on 1–2 January 2011

   Tables 5.11 and 5.12 compiled the premise and consequent parameters for eight steps head of ZTD prediction at NTUS using ANFIS FSC model. In this optimization, we found five rules for the best prediction of ZTD. The more the rules found in optimization (trial-and-error process) are likely to be more accurate in its

Table 5.9 The premise of ANFIS FCM Model $t + 8$ for NTUS station

Rule	if—part[a]																	
	$P(t - 1 s)$		$P(t)$		$T(t - 1 s)$		$T(t)$		$H(t - 1 s)$		$H(t)$	
	a	c	a	c	a	c	a	c	a	C	a	C
1	2.203	1004	2.231	1004	1.250	24.950	1.211	24.540	4.257	92.500	1.979	92.200
2	3.030	1010	3.035	1005	1.901	35.060	0.837	35.320	4.598	51.120	2.408	50.790
3	3.157	1007	3.172	1007	4.901	28.470	4.724	28.470	4.790	72.700	2.610	72.800

[a]{a, c} are gauss membership function obtained using Eq. 5.4

Table 5.10 The consequence parameter of ANFIS FCM Model $t + 8$ for NTUS station

Rule	Then—part
	Equation for f = consequent parameter
1	$-9.418 \times 10^{-3}P(t - 1\ \text{s}) + 1.168 \times 10^{-2}P(t) + 1.503 \times 10^{-3}T(t - 1\ \text{s}) + 1.289 \times 10^{-2}$ $T(t) + 2.686 \times 10^{-4}H(t - 1\ \text{s}) + 2.428 \times 10^{-2}H(t) + -0.293$
2	$7.781 \times 10^{-3}P(t - 1\ \text{s}) + -5.596 \times 10^{-3}P(t) + 1.355 \times 10^{-2}T(t - 1\ \text{s}) + -1.675 \times 10^{-3}$ $T(t) + 1.016 \times 10^{-3}H(t - 1\ \text{s}) + 3.908 \times 10^{-3}H(t) + -0.300$
3	$-7.170 \times 10^{-3}P(t - 1\ \text{s}) + 9.118 \times 10^{-3}P(t) + 6.440 \times 10^{-3}T(t - 1\ \text{s}) + 6.741 \times 10^{-3}$ $T(t) + 1.307 \times 10^{-3}H(t - 1\ \text{s}) + 2.331 \times 10^{-3}H(t) + -0.02684$

predictions but, with consequences, require high computing time. In contrast, the more the rule obtained does not guarantee the obtained ZTD values as accurate. This is because each process requires the adjustment of rules and membership functions to avoid overfitting. In other words, trial-and-error process is the best way to get how much the rule required for suitable prediction of ZTD.

Figure 5.13 shows the result of ZTD prediction at $t + 8$ ahead for NUTS station in UTC time. The figure also shows the prediction after $t + 8$ until 24 h. The result shows the ZTD in NTUS that is more tidy and smooth compared to ZTD trend at SYOG. In this case, the ANFIS models were successfully and completely follow the pattern of GPS ZTD. This indicates that ANFIS model developed with a one-min resolution is more accurate to predict the ZTD value. The different between GPS ZTD and ZTD from ANFIS is below 1 %, which implies that the highest of resolution data to be used in the prediction process will obtain a high accuracy of the ZTD value.

The overall assessment showed that the ANFIS models (FCM and FSC) are capable to estimate and predict the ZTD value. Both models are proven to be most reliable fuzzy clustering and ability to deal with uncertainties and noise. The MFs of type (Gauss) for all input variables and linear for output were successfully demonstrated the best prediction of ZTD. In this method, the number of MFs assigned to each input variable is chosen by trial-and-error. A good agreement has been obtained in the predicted values compared with the measurable values. These results indicate that the ANFIS models with low costs are a reliable and simple model for predicting the ZTD value with high accuracy. This implies that the ANFIS techniques offer an alternative approach to conventional techniques and can serve as reliable and simple predictive tools for the prediction. In other words, ANFIS model in this study is an alternative technique to estimate and predict the ZTD value by only using meteorological data as input.

Table 5.11 The premise of ANFIS FSC model $t + 8$ for NTUS station

Rule	if—part[a]											
	$P(t-1\ s)$		$P(t)$		$T(t-1\ s)$		$T(t)$		$H(t-1\ s)$		$H(t)$	
	a	c	a	c	a	c	a	c	a	c	a	C
1	−17.98	1013	25.53	1006	4.192	29.080	4.097	28.930	3.595	93.420	3.895	92.26
2	−12.24	1007	−8.426	1006	6.346	35.060	6.896	37.950	9.518	51.160	9.571	51.32
3	5.032	1006	5.023	1006	4.014	28.760	3.935	28.710	7.953	75.640	10.23	76.71
4	11.50	1010	11.57	1009	7.699	32.420	7.779	32.500	7.610	60.710	7.599	60.87
5	8.80	1004	8.611	1004	−5.319	18.750	−6.41	19.930	7.599	60.870	10.62	93.19

[a]$\{a, c\}$ are gauss membership function obtained from Eq. 5.4

Table 5.12 The consequence parameter of ANFIS FCM Model $t + 8$ for NTUS station

Rule	Then—part
	Equation for f = consequent parameter
1	$5.647 \times 10^{-3}P(t-1\ s) + -3.594 \times 10^{-3}P(t) + 2.551 \times 10^{-2}T(t-1\ s) + -1.067 \times 10^{-2}$ $T(t) + -1.449 \times 10^{-4}H(t-1\ s) + 5.561 \times 10^{-3}H(t) + -0.248$
2	$8.852 \times 10^{-3}P(t-1\ s) + -6.691 \times 10^{-3}P(t) + 1.182 \times 10^{-2}T(t-1\ s) + -3.668 \times 10^{-4}$ $T(t) + 1.143 \times 10^{-3}H(t-1\ s) + 3.996 \times 10^{-3}H(t) + -0.270$
3	$-7.941 \times 10^{-3}P(t-1\ s) + 9.940 \times 10^{-3}P(t) + 4.637 \times 10^{-3}T(t-1\ s) + 8.895 \times 10^{-3}$ $T(t) + 1.853 \times 10^{-3}H(t-1\ s) + 1.706 \times 10^{-3}H(t) + -0.082$
4	$3.665 \times 10^{-3}P(t-1\ s) + -1.553 \times 10^{-3}P(t) + 1.523 \times 10^{-2}T(t-1\ s) + -2.693 \times 10^{-3}$ $T(t) + 8.351 \times 10^{-4}H(t-1\ s) + 3.531 \times 10^{-3}H(t) + -0.220$
5	$1.924 \times 10^{-2}P(t-1\ s) + 2.138 \times 10^{-2}P(t) + -1.068 \times 10^{-2}T(t-1\ s) + 2.403 \times 10^{-3}$ $T(t) + 1.996 \times 10^{-4}H(t-1\ s) + 2.285 \times 10^{-3}H(t) + -0.123$

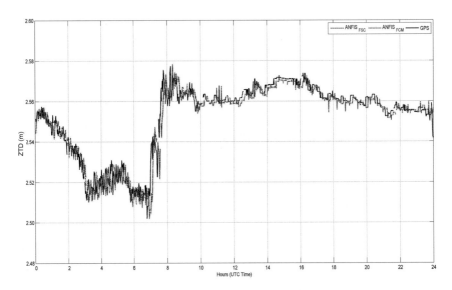

Fig. 5.13 The prediction of ZTD for NTUS station at $t + 8$ ahead with a one-min interval on 1 January 2010

References

Suparta W (2014) The development of GPS TroWav tool for atmospheric—terrestrial studies. J Phys: Conf Ser 495:012037. doi:10.1088/1742-6596/495/1/012037

Suparta W, Alhasa KM (2013a) Development of real-time precipitable water vapor monitoring system. In: Proceedings of 2013 3rd international conference on instrumentation, communications, information technology, and biomedical engineering (ICICI-BME), pp 135–140. doi:10.1109/ICICI-BME.2013.6698480

Suparta W, Alhasa KM (2013b) Application of ANFIS model for prediction of zenith tropospheric delay. Proceedings of 2013 3rd international conference on instrumentation, communications, information technology, and biomedical engineering (ICICI-BME), pp 172–177. doi:10.1109/ICICI-BME.2013.6698487

Suparta W, Alhasa KM (2015) Modeling of zenith path delay over Antarctica using an adaptive neuro fuzzy inference system technique. Expert Syst Appl 42(3):1050–1064

Suparta W, Abdul Rashid ZA, Mohd Ali MA, Yatim B, Fraser GJ (2008) Observations of Antarctic precipitable water vapor and its response to the solar activity based on GPS sensing. J Atmos Sol-Terr Phys 70:1419–1447

Index

© The Author(s) 2016
W. Suparta and K.M. Alhasa, *Modeling of Tropospheric Delays Using ANFIS*,
SpringerBriefs in Meteorology, DOI 10.1007/978-3-319-28437-8